REAPING THE REAL WHIRLWIND

REAPING THE REAL WHIRLWIND

A Biblical Response to the Theory of Man-Made Global Warming

DAVID I GUSTAFSON, PH.D.

VMI PUBLISHERS • SISTERS, OREGON

Reaping the Real Whirlwind
Copyright © 2008 by David I Gustafson, Ph.D.

All rights reserved

Printed in the United States of America
VMI Publishers, LLC, Camp Sherman, Oregon

ISBN: 1-933204-72-9
ISBN 13: 978-1-933204-72-7
Library of Congress Control Number: 2008933071

Library of Congress Cataloging-in-Publication Data
Gustafson, David I, 1958–
Reaping the Real Whirlwind: a Biblical Response to the Theory of Man-Made Global Warming / David I Gustafson

Unless otherwise indicated, all Bible quotations are taken from the New American Standard Bible. Copyright © 1960, 1962, 1963, 1968, 1971, 1972, 1973, 1975, 1977 by The Lockman Foundation, The Open Bible,® Copyright © 1978, 1979 by Thomas Nelson Publishers, used by permission.
Three Bible quotations, marked KJV, are from the King James Bible.

Cover design by Joe Bailen.

Contents

Chapter One	Modern Science vs. the Bible	9
Chapter Two	Climate Science	29
Chapter Three	Climate Change in the Bible	59
Chapter Four	Predicted Climate of the Twenty-First Century	97
Chapter Five	Predicted Climate Change Impacts	119
Chapter Six	"What Sort of People Ought You to Be?"	131
Chapter Seven	Recap and Benediction	147
Appendix 1	Annotated Bibliography	153
Appendix 2	Contact Information for the Author	176

List of Illustrations

Figure 2.1

Recent Increases in Atmospheric Carbon Dioxide 56

Figure 2.2

Prehistoric Temperatures and Atmospheric Carbon Dioxide 56

Figure 2.3

"Phase Space" Diagram of the Earth's Climate for the
Past 65 Million Years 57

Figure 2.4

Global Land Surface Warming Pattern of the Past Twenty Years 57

Figure 3.1

Growth of World Population and Atmospheric Greenhouse Gases . . 96

Figure 3.2

End-Times Events Described in the Biblical Book of Revelation 96

Figure 4.1

Actual Northern Hemisphere Land Warming vs. IPCC Modeling . . 117

Figure 4.2

Quadratic Fit to Recent Northern Hemisphere Land Temperatures . 117

Figure 4.3

Comparison of IPCC Models and Quadratic Fit Through
the Twenty-First Century 118

Figure 5.1

IPCC Summary of Expected Warming Impacts during
the Twenty-First Century 129

To
ישׂוה
and may His Holy Name be forever praised

Chapter One

MODERN SCIENCE VS. THE BIBLE

FIRST OF ALL, THANK YOU KIND READER for taking the time to read this book. I can't promise that these words will necessarily change your mind or change your life, but I'm pretty sure you will hear a perspective on the theory of man-made global warming you have not previously heard. The perspective is unique, because I will draw on both modern science and an ancient text, the Bible, in my sincere attempt to provide an objective discussion of whether the theory of man-made global warming is actually true.

THIS BOOK'S OVERALL PURPOSE

My overall intention in writing this book is to raise awareness among both born-again Christians and the secular world to a key conclusion that I have reached in my attempt to evaluate the theory of man-made global warming in light of what I believe to be the God-inspired truth contained in the Bible:

- *The theory of man-made global warming is entirely consistent with the Bible, and it is actually foretold by Scripture in surprising detail.*

Readers interested in joining a moderated debate about this statement and its implications are invited to participate in a continuing online forum on this subject being hosted at the following Internet web-site: http://www.real-whirlwind.org.

It is my hope that this book and the debate that it is likely to engender will ultimately lead to an expansion of God's Kingdom in at least three ways:

1. By correcting the science of those leading evangelical Christian organizations that are currently wrong on this issue, it will remove one objection among environmentally-minded youths as they come of age and are exposed to the gospel and those who profess it.
2. This ministry should also help win converts among current secularists who are absolutely convinced about the reality of man-made global warming but who have never considered the possibility that the Bible actually contains reliable prophecy through God's inspiration.
3. Finally, it should help remove a general objection to the gospel among secular scientists and those who follow their fundamental naturalist tenets, by forcefully defending the consistency of the Bible with all of modern science, not just this particular theory.

WHENCE THE BOOK TITLE?

The book title is taken from a frequently quoted passage in the ancient biblical book of Hosea, who wrote of the northern kingdom of Israel some 2,700 years ago, "they sow the wind, and they reap the whirlwind" (Hosea 8:7). In the case of man-made global warming, this law of the harvest is plainly at work. For what modern man is now doing may be simply understood as "sowing to the air"—an almost unimaginable quantity of stored photochemical energy in the form of carbon dioxide and numerous other powerful greenhouse gases. These annual global emissions represent uncounted millennia of trapped solar energy, and the amount we release each year continues to grow exponentially. According to the theory of man-made global warming, man is sowing this stored photochemical energy to the air and is now beginning to reap a *real* whirlwind of rising global temperatures and numerous other environmental calamities, many of which will be discussed within these pages.

This is simply the law of the harvest that God has made plain in all of Creation. As Charles Stanley often says, "you reap what you sow, more than you sow, later than you sow." It is just as the Apostle Paul warned us all in his letter to the Galatians nearly 2,000 years ago, "Do not be deceived, God is not mocked; for whatever a man sows, this he will also reap" (Gal. 6:7). Several of the gases humanity is releasing to the atmosphere have molecular structures that are invisible and transparent in gaseous form, permitting solar radiation to pass through the air, but they absorb infrared radiation of the kind that the Earth would otherwise emit back towards space. Although a small greenhouse effect is necessary for maintaining an environment suitable for life, an out-of-control greenhouse effect could create a planet completely inhospitable to life, such as on Venus, now bathed in an atmosphere of mainly carbon dioxide at around 867°F (464°C). Are we headed there? Only God knows! Can we do anything about it? Read on!

Snapshot of the Author

But before going too much further, I'd like to take a brief moment to introduce myself. I grew up in the Pacific Northwest of the United States, where I developed a profound love for the outdoors. Our family of six camped each summer weekend at a lake in northeastern Washington, where only fly-fishing and no motor boats were permitted. As a youth, my "spiritual" home life included exactly two visits to Sunday school when I was eight years old and annual celebrations of Easter and Christmas always with the Easter Bunny and Santa Claus. These holidays were wildly elaborate, very fun for us as kids, but entirely secular with no explanation at all about the "reason for the season."

Following my Dad's lead, I grew up ridiculing Christians, especially my classmates in high school. I still have vivid memories of a debate as part of my advanced high school biology class where I took on a fellow senior, Stewart Sonneland, concerning the question of Creation vs. godless Darwinian evolution, the latter of which I fiercely defended as the only logical possibility. Interestingly, although my parents were both "devout" agnostic/atheists and dismissed all of Christianity as a spiritual crutch for unsophisticated mental midgets, they did purchase and regularly listen to the double album "Jesus Christ Superstar," which provided my only real childhood exposure to the "gospel," albeit a quite distorted one (as I now have come to realize).

Following high school graduation and one last summer of weekend campouts at the lake, my parents delivered me to Stanford, where I received a B.S. in chemical engineering after four years of an intermittently studious term of service. Upon graduation in 1980, I turned down various job offers to enter one of the industries hiring chemical engineers at that time: aerospace, consumer products, nuclear, and oil. Instead, I moved back to the Pacific Northwest, where I began pursuing a doctorate in chemical engineering at the University of Washington in Seattle. Soon after the fall quarter began, I became an avid runner (which was also a hobby of my father), and I successfully ran the Seattle Marathon on the Saturday after Thanksgiving, 1980.

About a month later, on 26 December 1980, my previous life of blissful ignorance of all things biblical suddenly came crashing down when I experienced a sudden conversion experience during one of my daily training runs. Unlike Paul on his way to Damascus, I was not felled to the ground by a blinding light, but, like the apostle, I was completely overwhelmed by an immediate and irresistible need to completely surrender my life to Jesus. As soon as I came home, I began voraciously reading the Bible, and I began witnessing to my immediate family (siblings and parents), all of whom were convinced I had gone completely crazy. To this day, only a few members of my entire extended family of origin have confessed Jesus as Lord. This is a daily source of prayer for me, as it has been for nearly three decades now.

Anyone who has tried living the Christian life knows it is not easy, particularly in today's world, and especially within a family where one is so badly outnumbered and ridiculed. But the Lord has been gracious to me, despite how many times I have fallen away and off the narrow path He has predestined for me to follow. I won't "bore" the reader with all the gory details. I am a sinner, but I thank God I'm not the man I used to be! And I know He loves me too much to leave me the way I am, up to the very moment I am sharing this testimony.

I then went on to successfully defend my doctoral dissertation on 26 September 1983, the day before my twenty-fifth birthday, and became employed in the agricultural industry where I have been a practicing environmental scientist ever since. During my professional career I have previously authored a scientific textbook, *Pesticides in Drinking Water (1993)*, and nearly sixty peer-reviewed research papers in my area of specialization—the use of computer simulation models to address environmental issues in agricultural systems. In 2003, my employer named me a Senior Fellow, a rank reserved for scientists who have reached a level of technical contribution that has extended broadly inside and outside of the company.

Because the question of global climate change has become such a politically-charged issue, I believe it is also important for me to briefly disclose my political beliefs. Consistent with my family of origin, I arrived at Stanford as a liberal Democrat, but I graduated as a conservative Republican. This abrupt change horrified my parents, who I am sure rued the day they ever sent me to Stanford.

After leaving academia and through most of the subsequent twenty years of my work life, I attempted to settle into the comfort of American suburbia, seeking the "good life," and succumbing to the consumerism that so dominates much of the United States. However, beginning with my father's year-long battle against lung cancer, and intensifying after his death in August 2000, I found myself growing more deeply concerned with matters of eternal consequence. I became a regular listener of several Christian radio broadcasters on the Bott Radio Network, carried by KSIV in St. Louis, including: Alistair Begg, Ravi Zacharias, Charles Stanley, Chuck Colson, Chuck Swindoll, Chip Ingram, Woodrow Kroll, R. C. Sproul, James MacDonald, John MacArthur, James Dobson, David Jeremiah, Hank Hanegraaff, Steve Brown, and the now-deceased trio of D. James Kennedy, Adrian Rogers, and J. Vernon McGee. I also began playing guitar in the praise band of a nearby neighborhood church each Sunday, and I began studying the Bible again with some of the same fervor and passion that I had exhibited in the first few years of my conversion.

It was from this perspective that I first encountered the theory of man-made global warming. My initial response to the theory was extremely negative, largely because I associated it almost entirely with its most noticeable proponent, Al Gore. I had long before dismissed Mr. Gore as an opportunistic politician who was quick to adopt positions out of expediency, rather than conviction, as evidenced by his switching from a pro-life to pro-abortion position when leaving Tennessee state politics for competition at the national level. In addition to opposing the theory on mere political grounds, I had scientific and theoretical objections as well.

During the late 1980s I had read a then popular book, *Chaos,* by James Gleick, who introduced me to the seminal work of an MIT scientist, Konrad Lorenz. Dr. Lorenz had discovered, first empirically, and then subsequently proved theoretically, that certain types of mathematical equations (for the mathematically inquisitive, coupled systems of non-linear differential equations), such as those that govern weather, are intrinsically unstable numerically. This discovery, later given a popular name, the "Butterfly Effect," imposes profound limitations on the science of weather forecasting. By the way, the name comes from the seemingly absurd possibility that if a butterfly flaps its wings the wrong way in Peru one afternoon, it could possibly cause a snowstorm in Peoria next week. This led me to the following improper inference: "If Lorenz proved you can't predict the weather two weeks from now, then it's ridiculous to believe computer models could accurately predict the climate 100 years from now."

What I didn't understand at the time I reached that faulty conclusion, was something that I recently learned while spending an afternoon with Dr. Stephen Schneider in early 2007. He is the Stanford scientist who has played such a key role in developing and championing modern climate science. Some might say he "invented" it (which is far closer to the truth than Mr. Gore's notorious claim to have invented the Internet!). Dr. Schneider very kindly and patiently explained to me that the Lorenz Butterfly Effect is applicable only to Initial Value problems, such as the case of trying to predict the precise weather for the next few days based on a particular set of initial conditions—whereas predicting the long-term trend in climate is a Boundary Value problem, for which there is no such numerical stability limitation.

I know some of my readers are now getting worried about the level of mathematics in this discussion. As the angel Gabriel nearly always first exclaims, "Fear not!" For those lacking knowledge on the mathematics of differential equations—don't worry! That's the last time I will use these terms. But trust me, Dr. Schneider is right!

Now there are still very real limitations and uncertainties in the computer simulations used to predict long-term climate change trends, but the Butterfly Effect is not one of them. My interesting (and, may I add, providential) meeting with Dr. Schneider helped spawn within me a passion to study the theory of man-made global warming with great fervor. My study has involved a careful review of both the scientific literature and the entire Bible. I have been astonished to discover that the predictions of the climate scientists are eerily similar to the biblical descriptions of the Last Days. It is this amazing "coincidence" in the teachings of science and Scripture that has inspired this book. Ever since my work on this book project began, I have felt a continuing prompting of the Holy Spirit. I humbly pray that this work will be found acceptable to Him.

Why Another Global Warming Book?

Much has already been written on the theory of man-made global warming, but I have not found an example of the approach taken in this book. I believe it to be unique in the sense that I will employ both biblical texts and the latest work of the world's leading climate scientists in seeking an answer to the following deceptively simple, yet profoundly implicative question:

- *Is the theory of man-made global warming actually true?*

It is my sincere hope that this book will help raise awareness about the reality of man-made global warming to believers and nonbelievers alike. In so doing, I seek to provoke each one of us to respond in a wise manner to the stewardship challenge that global warming presents. This is the only Earth God gave us, and He clearly gave us dominion over it from the very beginning, "be fruitful and multiply, and fill the earth, and subdue it" (Gen. 1:28). The first job He gave the first member of our species (we know him as Adam) after He placed him in the Garden of Eden was "to cultivate it and to keep it" (Gen. 2:15). So now you know why gardening is such a relaxing and popular pastime! However, we need not look too far to see how profoundly we have failed to maintain and care for this planet in the manner that God had commanded.

As a born-again believer, I know that my primary calling is to go forth "and make disciples of all the nations" (Matt. 28:19). But it is my sincere belief that I am also now being called to awaken many of my Christian brothers and sisters from their slumber of indifference about the harm our actions are apparently causing to God's Creation. If the science is right, as I intend to show, we are destroying this planet as we continue to lead lives of thoughtless waste. As a follower of Jesus and a believer in the teachings of the Bible, I am called to love God with all of my heart, with all of my mind, and with all of my strength – and to love my neighbor as myself (Luke 10:27). I read nothing about loving my car, my electronic gizmos, and my next fancy meal. Though I may actually need at least some of these things to survive, I am not to love or worship them, for that would be idolatry. Idolatry is a word that has lost any meaning for most of the world, but I pray believers will find it within their hearts to ask the tough questions about where their true desires lie. We do not follow the Lord's first great commandment to love our Heavenly Father and our neighbors if we remain indifferent to the harm our acts of commission and omission may be causing.

This book is primarily intended for all Christians with an interest in the theory of man-made global warming. The science will be covered in such a way that it should be completely accessible to anyone with a high school level of science training. But I also intend for the book to be rigorous enough and replete with sufficient references for it to be used in appropriate academic settings.

To the reader interested in global warming, but who has not yet discovered that Jesus is indeed God, the "very God who spoke and the universe leapt into existence," (a great phrase from Hank Hanegraaff, "the Bible Answer Man") I invite you to follow along as I take you though my personal encounter with the Bible and this theory. Like you, I once dismissed the Bible as a compendium of myths and primitive tales, but if you are patient enough to hear me out, I think you will find that Scripture has a lot to say about the present hour in which we live—straddling a very precarious precipice—on the brink of global catastrophes of untold human suffering unprecedented since man first appeared on this planet, exceeding even the damage of the Flood. Or, as Jesus described it, "for then there will be a great tribulation, such as has not occurred since the beginning of the world until now, nor ever shall" (Matt. 24:21).

"A Nature Hike Through the Book of Revelations [sic.]"

In the Oscar-award-winning movie, *An Inconvenient Truth,* Al Gore is famously heard describing the predicted impacts of impending global climate change as being "like a nature hike through the Book of Revelations [sic]." For those of us who are born-again believers, many of whom happen to vehemently disagree with Mr. Gore on a number of political and moral issues, it is particularly grating on the ears to listen to him at all, but especially when he botches the title of this precious, final book of our Holy Scriptures. But I did watch his movie one and a half times, the second time just long enough to ensure that I have properly quoted him here.

Few readers can be unaware of our former vice-president's recent documentary and the book that spawned it. I found myself surprised by the extremely personal nature of both the book and the movie, but perhaps I shouldn't have been. A good bit of the science presented by Mr. Gore is either overly simplified or intentionally misleading (such as reversing the causal relationship between historic temperature and carbon dioxide levels), but there are some great visual aids in a largely successful effort to communicate the nature of what is currently happening to the world's climate. I congratulate Mr. Gore on his passion and dedication to the project. However, I'd suggest that he take a closer look at his personal rate of energy consumption, rather than flying around the world to lobby for ineffective treaties.

Yes, Al, I am throwing stones, and though I am not without sin, and I am sure there is some improperly insulated glass in my home, I feel compelled to offer a testimonial about my own carbon footprint. Since embracing the reality of global warming in early 2007, I've moved far closer to work in order to reduce my commute, and I now inhabit an 800 square foot apartment immediately next to a MetroLink (St. Louis light rail) stop. I've also ditched my minivan for a used 2003 *Prius* hybrid, reduced my consumption of animal-based foods, and I intentionally avoid all driving for at least a twenty-four hour period each weekend (a "Carbon Sabbath," which in this respect is similar to the conventional Sabbath celebrated weekly by our nearest spiritual neighbors, the Orthodox Jews). Small steps each, but cumulatively I estimate that my personal carbon footprint has decreased by more than half since making these changes. Can it make a real difference? I will save that discussion for Chapter 6 and instead get back to quibbling with Mr. Gore.

The part where I find myself most strongly disagreeing with the former Vice President is on the "so what?" question. Gore's take on this question is betrayed by the subtitle to his book, "what we can do about it." His answer is the typical leftist-socialist plan of massive governmental bureaucracies and global regulations at the expense of personal freedoms. In so doing, Gore and his secular environmentalist allies deny the biblical reality that the only entities in this present world that are truly eternal are the Word of God (The Bible) and the souls of every individual now placed upon this Earth. The Bible teaches that it is the enemy of our souls (Satan) that asks us to deny that reality in favor of worldly comfort. I see this clash of worldviews on whether the Creation or the Creator is to be worshipped as evidence of the global spiritual battle that is now raging between the forces led by Jesus and the forces of Satan. We know who wins, so I'd humbly suggest that you choose wisely whose side you are on. There is no middle ground.

LIMITS OF HUMAN UNDERSTANDING—
THE FINITE VS. AN INFINITE TRIUNE GOD

Before going any further on a topic as profound as this, I feel compelled to stress the limits of my own capacity to understand all that God has planned for us in this global predicament. Each of us should approach a subject such as man-made global warming with humility. There are limits to human understanding that are intrinsic to the fact that we are, and will forever remain, finite: "If anyone supposes that he knows anything, he has not yet known as he ought to know" (1 Cor. 8:2). However, the God we worship is infinite, omnipotent, and possesses a triune nature (Father, Son, and Holy Spirit). Indeed, this profound truth is displayed in the triune universe that He created: "In the beginning, God made the heavens and the earth" (Gen. 1:1)—time (beginning), space (heavens), and matter (earth). All three bear His triune nature: time has past, present, and future; space has length, width, and height; and matter has protons, neutrons, and electrons. These stable elementary particles are held together by three dominant forces at three increasing scales: nuclear, electromagnetic, and gravitational. If we shift attention from the external universe and examine ourselves, we also find evidence of the triune nature of the God in Whose image we have been created: body, soul, and spirit.

God's triune nature is not inconsistent with the fact that there is only one God. This is the monotheistic belief that was the cornerstone of Judaism and distinguished it from the surrounding polytheistic belief systems of its time of origin. Biblical Christianity, birthed directly from Judaism, fully retains this monotheistic faith—but the one God is understood to comprise three eternally distinct persons—known in theology as the doctrine of the trinity. The fact that we can not fully comprehend this truth, which is revealed only in Scripture, is not ultimately a reason to doubt the validity of Scripture but is rather a reassurance that the God we worship is infinite in extent. He is holy, completely separate from the Creation with an "otherness" that we must humbly acknowledge. He chose to create us and give us free will—knowing we would all fail to obey Him. He then sent His Son to take on human flesh in order to pay an infinite redeeming price—death––for that sin. He did all of this so that we would then be free to worship and enjoy Him forever. This supremely loving series of acts should cause us each to fall to our knees and humbly thank Him for all He has done. That even an hour can go by without me doing so myself is a true measure of the depravity I find in my heart.

Roadmap for the Remainder of Chapter 1

Having now given some of my personal background and confessed my own intellectual and spiritual limitations, I will take the remainder of this opening chapter to explain and defend my methodology and approach. A necessary portion of this opening chapter is devoted to my practice of freely alternating between science and Scripture, which I feel should be granted equal time. My intent is to be consistent as I examine the controversial theory of man-made global warming from the perspective of a practicing environmental scientist who happens to have both a scientific and a biblical worldview.

Wait a Minute!

Yes, I know there are some of you, especially those who have not yet investigated or accepted the claims of Jesus, who will immediately object to me "dragging religion" into this discussion of a "purely scientific" issue. I make no apologies for taking this approach, and I simply ask for your forbearance and willingness to keep an open mind until you have heard me out. I also welcome your feedback and any objections you may have, philosophical or otherwise, to be courteously expressed and posted on the web-site previously mentioned: http://www.real-whirlwind.org.

A Historical Perspective on Science and the Christian Faith

The reason many will object to this intentional juxtaposition of science and Scripture is that they have willingly accepted what has become a majority view in the largely secular societies of the United States and elsewhere that "religion" and "science" should be kept as entirely separate and distinct disciplines. In many parts of today's world, those having views informed by faith are asked to keep them as a private matter, not too different from the way we now treat smokers in much of modern society. By attempting to banish all Scripture-based opinions from the public marketplace of ideas––particularly college campuses—our modern world has long ago forgotten what was originally meant by the name of our places of higher education, "university."

This word, "university," is one whose etymology conveys the original idea of the founders of these institutions, which was to bring together "unity" and "diversity" in the minds of our most intellectually-gifted young people. The word was meant to convey the notion that there is an overarching unifying source, namely God, who is behind all of the various branches of knowledge and that such institutions were a perfect place for the most talented of our children to receive such instruction. Now, of course, modern campuses no longer live up to the lofty definition of the word, and they are little more than a place to get equipped for a high-paying job, party hardily, and perhaps meet a potential mate or two—or even more likely to just temporarily "hookup" to use the profane vernacular of our day.

But it was not so at the beginning of our nation's history. Institutions such as Harvard and Princeton were founded by Christians who believed that the study of both the sciences and the Bible were equally essential to the healthy intellectual development of well-rounded, properly balanced, and completely educated leaders of society. It is little wonder that individuals educated in such a way could author these famous words, "we hold these truths to be self-evident, that we are endowed by our Creator with certain unalienable rights..." Chief among the signers of the Declaration was John Witherspoon, then president of Princeton and an ordained minister.

Clearly, opinions in the United States have changed from those of our Founding Fathers, and we no longer generally accept the idea of scientists and biblical scholars speaking with equal authority on the same subject. But is there a rational basis for completely excluding the idea of examining scientific theories from a biblical perspective? I answer no, and I first cite the faith, largely a Christian faith, among most of the world's greatest founders of modern science. Sir Isaac Newton (1642–1727) is undeniably first among this list. He was recorded to have said, "I have a fundamental belief in the Bible as the Word of God, written by men who were inspired. I study the Bible daily." As the developer or codeveloper of so many of the modern sciences: gravitation, calculus, and the laws of motion, to name but three, Newton stands above the rest. But he is not alone. One particularly compelling way to understand this point is to simply consider the long list of the branches of modern science founded by other Christians. It is impressive:

- Roger Bacon (ca 1220–1292), the scientific method itself
- Johannes Kepler (1571–1630), the laws of planetary motion
- Blaise Pascal (1623–1662), mathematics and philosophy
- Robert Boyle (1627–1691), modern chemistry
- Antonie van Leeuwenhoek (1632–1723), discoverer of bacteria
- Carolus Linnaeus (1707–1778), modern biological classification
- Leonhard Euler (1707–1783), higher mathematics
- John Dalton (1766–1844), modern atomic theory
- Michael Faraday (1791–1867), electrical phenomena
- John Frederick William Herschel (1792–1871), astronomer

- James Prescott Joule (1818–1889), physics
- Gregor Mendel (1822–1884), genetics
- William Thomson (*aka* Lord) Kelvin (1824–1907), thermodynamics
- James Clerk Maxwell (1831–1879), electromagnetism
- George Washington Carver (c. 1864–1943), chemistry of biological materials

The Royal Society of London, which many would consider to be the world's premier scientific organization, certainly from a historical perspective, was founded in the seventeenth century by ten scientists, all Christians––seven of whom were Puritans!—and all ten had a biblical view of Creation.

In our day, many practitioners and leaders of modern science also happen to be Christians. Among the most prominent of these is Francis S. Collins, past head of the Human Genome Project. Collins is a self-proclaimed born-again believer, and he sees complete compatibility between modern science and the Christian faith. Even Einstein, of Jewish heritage by birth, is famously remembered to have questioned the underlying basis of quantum mechanics with the statement, "God doesn't play dice." The entertaining retort to this statement is less well remembered, but it came from a Nobel-winning contemporary of Einstein's, Neils Bohr, who boldly confronted him, "Einstein, quit telling God what to do!" Interestingly, despite Einstein's Jewish family of origin, he was convinced of the historicity of Jesus, and is recorded in Isaacson's recent biography (2006) to have said, "no one can read the Gospels without feeling the actual presence of Jesus. His personality pulsates in every word. No myth is filled with such life" (*Einstein* p. 386).

Another powerful argument for why it is reasonable to permit biblical analysis of scientific theories goes back to an objective reexamination of the nature of what really defines "science." In our day, a majority of practicing scientists may be defined as having a "naturalist" philosophical perspective. A naturalist refuses to accept the possibility of any supernatural causes having contributed to the observable universe. The basic assertion of the naturalist is as follows:

> The entire universe can be described through a discoverable set of comprehensible mathematical and physical laws that may be

empirically confirmed by the collection and interpretation of experimental data. No "supernatural" explanations are permitted.

A classic example of this thinking is the debate over biological origins that today pits Darwinian evolutionists against the adherents of Intelligent Design. The naturalist stubbornly clings to the crumbling theory of chance and unguided natural selection as somehow giving rise to the mind-boggling complexity of living organisms, despite the ever-growing mountain of experimental evidence that clearly points to an unimaginably intelligent and powerful Designer. By adamantly refusing to permit any such supernatural explanations, the naturalist reveals an arrogance that I believe has its ultimate roots in human pride.

The overarching guiding principle of the naturalist is that all supernatural explanations must be excluded. This is itself an improvable supposition—an article of faith—and it is the entire philosophical basis of the naturalist. Notice that I did not say this is the underlying philosophy of the scientist. Contrary to the scientist, the naturalist dismisses possible explanations—the very antithesis of the open objectivity that should characterize the mind of an intellectually honest scientist. Just like the many scientists mentioned earlier in this chapter, I am one who remains open to the possibility that some of what we observe in the universe can not be explained through a tidy set of physical laws operating under the following silly syllogism: "nothing suddenly became everything, and once it endured a sufficient number of undirected random events, it became the wonderfully ordered and information-rich universe we now observe." I find such thinking to be patently absurd and inconsistent with what we now know to be true, from molecular biology and all the modern sciences. King David wrote some three thousand years ago, "The fool says in his heart, 'There is no God'" (Ps. 14:1).

The Biggest "Scientific" Objection to the Bible: Six Days of Creation?

Having made the case, I hope, for the intellectual legitimacy of the approach that I will be taking in this book, I'd like to address the single biggest objection that I hear when defending a biblical worldview before scientists. It's the very beginning of the Bible that appears to unequivocally present the creation of the universe as having been accomplished in six twenty-four-hour days. It is a huge stumbling block for many, both scientists and non-scientists alike. From early in our childhood, we now learn many facts that seem to scream out against the Genesis account: hundreds of millions of years of dinosaurs, the roughly 4.6 billion year age of the Earth, and the roughly fourteen billion year age of the entire universe. But I personally believe the secret of reconciliation lies in the work of a man already mentioned earlier in this chapter, Einstein. For this explanation I am indebted to the writings of Gerald Schroeder, who is both a theoretical physicist and an Orthodox Jew with a very high view of the authority of Scripture and author of the compelling book, *Genesis and the Big Bang,* published in 1990. As Schroeder reminds us, Einstein convincingly demonstrated that time is not absolute and that it is instead entirely dependent on our inertial frame of reference (how fast we are moving). Schroeder goes on to present the brilliant hypothesis that the "six days of creation" should be measured from God's frame of reference, from which they could be properly construed as "actual" twenty-four-hour days. A line from Peter comes to mind, paraphrased from the only Psalm ascribed to Moses, (90), "with the Lord one day is as a thousand years" (2 Pet. 3:8). According to Schroeder, Peter's and Moses' hyperbole is not quite as hyperbolic as it should be, and the days would actually correspond to a decreasing geometric series measured in billions of years (approximately 8, 4, 2, 1, ½, ¼)—values which Schroeder argues are consistent with both astronomical physics and the geological record of this planet.

Genetic and anthropological data appear to be in agreement that man appeared in Africa somewhat more than 100,000 years ago (measured from the Earth's inertial frame of reference). Modern human genetic data are entirely consistent with the hypothesis that every member of our species arose from a single pair of individuals. The Genesis account gives these two individuals very famous Hebrew names meaning "dust" (Adam) and "living" (Eve). I challenge my Darwinian evolutionist neighbors to explain how this first couple's children were able to successfully reproduce without suffering the immediately deleterious effects of recessive mutations, which today render most progeny of close relatives quite malformed. The only logical answer is that they had not yet accumulated the multiplicity of harmful mutations that are now found in our genomes. Even Abraham (ten named generations after Noah) is recorded to have married his half sister, Sarah, who gave birth to Isaac, the father of Jacob (later renamed Israel), the patriarch of all of today's Jewish people. The combination of fewer mutations and lower population density apparently made this type of a marriage both practicable and necessary.

However, I fear that I now may have offended the sensibilities of many of my Christian brethren who vigorously defend the idea of a "young Earth." They question the reliability of the radioactive isotopic dating methods that are now in widespread use throughout the scientific community—and which appear to be entirely self-consistent according to all that I have read. To the "young Earthers" among those still reading this book, I would cite Hank Hanegraaff's oft-repeated phrase, "this should not be an issue we divide over."

Source Documents

In assembling the information I will discuss in this book, I have relied upon a number of recent books, reports, and articles published in the scientific literature and the popular media. When these documents are cited in the text, I will give the name and author in the conventional parenthetical form (Name, year), unless the author is mentioned in the sentence, when I will provide only the year. These reference documents are listed as an annotated bibliography in Appendix 1. In giving this list of references, I do not mean to imply that it represents an exhaustive search of the literature on the topic. Indeed, the body of written material on this subject has grown so rapidly in recent years that I don't believe it would be possible for even a team of researchers to adequately summarize all of it. To paraphrase the familiar hyperbole of John at the end of his gospel, "I suppose that even the world itself would not have room for the books that would need to be written" (John 21:25). However, I offer the list and my comments on the contributions by each of the authors in a sincere attempt to point the interested reader to some particularly useful articles and texts that I have studied while preparing this book.

Of course, the task of assimilating and synthesizing all of the known science about the theory of man-made global warming and its possible impacts on humanity is exactly what has been taken on by the Intergovernmental Panel on Climate Change (IPCC), as convened and supported under the auspices of the United Nations. And I will therefore rely heavily on IPCC documents as I present the science—though I will give ample attention to the many critics of IPCC as well.

As for the Biblical references that are scattered throughout this book, I will rely (unless otherwise noted) on the New American Standard translation provided in "The Open Bible," published by Thomas Nelson Publishers in 1979.

Outline for the Book

Here is a brief outline for the remaining chapters of the book. In Chapter 2, "Climate Science," I describe the history of modern climate science, with its humble beginnings of the early twentieth century, followed by the more recent dramatic advances made possible by the advent of supercomputers. The discussion begins with the story of the Swedish scientist, Svante Arrhenius, who first proposed (in 1896) the theory that carbon dioxide emissions from burning coal would warm the atmosphere. The chapter ends with objective consideration of the stated concerns of the many "contrarian" scientists who have challenged or otherwise questioned the IPCC consensus-based conclusions.

In Chapter 3, "Climate Change in the Bible," I take the reader through a comprehensive search and discussion of biblical references to the Earth and its climate, highlighting man's contributory role toward its disruption and ultimate destruction. From the first book to the last, the Bible teaches that God placed the Earth under the dominion of man, but we have supremely failed to live up to that obligation. For that crime, God will allow us to take this planet to the brink of complete destruction, until He intervenes one last time in human history to end the madness. He will rescue those who love Him and are still alive, to spend eternity with Him in a new heavenly home—but He will permit the remainder to spend eternity forever separated from His loving presence.

Chapter 4 is entitled, "Predicted Climate of the Twenty-First Century." It summarizes the key climate predictions for the remainder of this century, as described in the most recent set of IPCC reports. I will present a simple empirical analysis of the available temperature data that suggests current climate models are actually too sluggish, with observed temperatures rising at about twice the rate predicted by IPCC.

In Chapter 5, "Predicted Climate Change Impacts," I rely mainly on the IPCC reports to describe the impacts now predicted to take place later in this century as a result of climate change, including widespread species extinctions, human disease, and famine. A particular emphasis is placed on the recently discovered phenomenon of "ocean acidification," which is a direct result of the absorption of atmospheric carbon dioxide by the world's oceans. This change in oceanic chemistry will lead to the dissolution of calcium carbonate, a key component of numerous marine organisms, which will further threaten food webs and fisheries already crippled by overfishing.

The title of Chapter 6 comes from a rhetorical question asked by the apostle Peter, "what sort of people ought you to be" (2 Pet. 3:11). It seeks to answer the "so what?" question from the perspective of one with a biblical and scientifically-informed worldview. When Peter asked the rhetorical question of this chapter title, his answer was that we ought to live lives of "holy conduct and godliness, looking for and hastening the coming day of God" (2 Pet.3:11–12). The chapter seeks to bring this admonition up to date with regard to our God-given dominion over the Earth, the commandment to love our neighbors as ourselves, the Great Commission, the scientific limits to what can actually be accomplished to slow the current warming and other man-made impacts, and the tension between God's sovereignty and man's free will.

As suggested by its title, "Recap and Benediction," Chapter 7 is intended to briefly reprise the previous discussions and wish the departing reader well, despite the horrendous weather forecast for the planet. It is followed by the annotated bibliography of Appendix 1, which gives the serious reader a series of references and web-site links to more detailed information. But now it is time to look at the underlying science.

Chapter Two

CLIMATE SCIENCE

I'M JUST OLD ENOUGH to have taken high school chemistry when we were all forced to learn how to use slide rules, and no students were permitted to use handheld calculators, which were just becoming widespread. In fact, I even had a science fair project where I designed and built a working circular slide rule that had four scales and enabled rapid numerical solutions to quadratic equations. But my timing with this invention was not exactly Edisonian. Although it may have won the local science fair, it was made completely obsolete by the computer age. When I entered Stanford in the fall of 1976, I took a beginning class in Computer Science, which at the time meant keypunching cards and feeding them into card readers. It was the very next quarter, in January of 1977, when the first computer time sharing system, LOTS, was made available to Stanford's byte-hungry undergraduate community. A few years later, in late 1984, I purchased a "portable" computer from Compaq, which was about the size of my mom's sewing machine. It had no hard drive, only 256 kilobytes of RAM, and a tiny green CRT. However, it had far greater longevity than all of my other PC's, probably because it never suffered the slow death of being connected to the Internet, which now seems to infect and eventually hobble virtually every machine that it touches.

Of course, that same Internet is a key enabling technology in the information age that has now transformed so much of commerce and research. On a parallel path, the sheer computational power of our personal and "mainframe" computers has been exploding at a breathtaking pace, delivering simulation capabilities that seemed wholly inaccessible only a few years ago. Along with this vast computing power, voluminous quantities of telemetry are now available monitoring the condition of the planet—essentially instantaneously—from anywhere, to anywhere, 24-7-365, to anyone with the technology to receive it. This has resulted in rapid increases in the ability of climate scientists to calibrate and otherwise improve their models.

In this chapter, I'll attempt to briefly summarize the development of modern climate science. Most of the early portions of this history will rely heavily on the excellent and well-written book by Spencer H. Weart, *The Discovery of Global Warming*, published in 2003. He explains the history of some of the early discoveries relevant to the theory that were made by a handful of scientists—including some recognizable names to those of us in the larger scientific community but also others that have remained cloaked in general anonymity.

Humble Beginnings

The story begins early in the nineteenth century with a French scientist named Joseph Fourier, who attempted to calculate the theoretical temperature of the Earth using the new physics of "blackbody radiation." This law states that warm bodies radiate heat in proportion to the fourth power of the difference in temperatures between the body and its immediate environment. Unfortunately, using this law, Fourier found that the temperature for Earth came out well below freezing (−18°C or 0°F) instead of the observed value of about 15°C (or 59°F and rising!). Fourier recognized that the atmosphere must somehow trap some of the heat contained in this outgoing infrared radiation.

The answer to this puzzle came in 1859 from a British scientist, John Tyndall, who began measuring the ability of various atmospheric gases to absorb such infrared radiation. He found that the two most common components of the atmosphere (oxygen and nitrogen) do not absorb it; however, he also tested "coal gas," which we now know is largely methane, and found that it does absorb infrared radiation. Methane, by the way, is currently ranked number two on the list of man-made greenhouse gases in terms of its current warming influence. Tyndall went on to try other gases, including water vapor and carbon dioxide, both of which he correctly found to absorb infrared radiation. He knew that these gases were present in the air at extremely low levels, certainly relative to both nitrogen (79%) and oxygen (20%), but he nevertheless concluded that they could help warm the planet, "as a dam built across a river causes a local deepening of the stream, so our atmosphere, thrown as a barrier across the terrestrial [infrared] rays, produces a local heightening of the temperature at the Earth's surface."

The first scientist to attempt detailed calculations on how much temperature change would be caused by variations in the atmospheric concentration of carbon dioxide was a Swede, Svante Arrhenius, whose last name should be recognizable to college-level physical chemistry students, for the reaction rate vs. temperature plots that still bear his name. During 1896, allegedly in the melancholy boredom of having lost his wife and son to divorce, he spent month after month attempting the global calculations, which supercomputers now perform in nanoseconds. The numbers he got were not bad given the limitations of the time, and though they were not precise by today's standards, they were dutifully reported in the scientific literature of the day. He began conversing about them with colleagues, including another Swede, Arvid Högboom. Although Arrhenius had started the calculations in an attempt to explain the occurrence of ice ages, Högboom brought the calculations up to more immediate relevance when he noticed that current global releases of carbon dioxide by factories and other human sources at that time were of similar magnitude to the levels that might be released by volcanoes or other natural events. Given the very low amounts of fossil fuel burning at that time, neither Arrhenius nor Högboom was particularly alarmed by the results, and a little warming sounded nice in Sweden anyhow. But the main reason for their lack of concern was that Högboom incorrectly assumed it would take millennia for human activities to double the amount of carbon dioxide in the air, thereby warming the Earth's temperature by 5–6 °C (10°F), as suggested by his calculations. But, of course, we now know that Henry Ford had other plans!

Twentieth Century Skepticism

Subsequent scientific scrutiny of Arrhenius's and Högboom's calculations brought a large degree of skepticism in the early twentieth century. Some of this was based on flawed experiments and flawed interpretation of those results, but a large degree of the skepticism came from a prevailing view that the "Balance of Nature" would find a way to respond in such a manner so as to counteract any potential disturbances to the system. This idea of "stationarity" is a cornerstone in the engineering design of levees and other flood control measures, where it is assumed that the statistical frequency of stream-flows will continue unchanged through time (e.g. a so-called "500-year" flood). This mindset is still very prevalent among scientists of today that stubbornly refuse to accept the possibility of human influences on climate.

So it was into this skeptical "climate" that an audacious young scientist, Guy Stewart Callendar, bravely stepped before the Royal Meteorological Society in 1938 to announce his new theory of man-made global warming––that carbon dioxide emissions were warming the Earth. His claim was made all the more brash by the fact that he was not a professional meteorologist, only an engineer who worked on steam power. However, he had carefully assembled global temperature data and information on atmospheric carbon dioxide levels to support his bold theory. As with Arrhenius, a fellow northern European, Callendar believed that a little warming would be a good thing, perhaps even helping agriculture. And, just like Arrhenius, he incorrectly prognosticated a very gradual increase due to his assumption of a very slight rise in atmospheric carbon dioxide, and too weak of a dependence of global mean temperature on this parameter. But his hypothesis was largely ignored as just another half-baked idea from a scientific crackpot, to be filed somewhere alongside…perhaps, another bogus claimed invention of a perpetual motion machine.

It was also during the 1930s that a Serbian engineer, Milutin Milankovitch, carried out excessively difficult and tedious calculations involving slight variations in the Earth's orbit, which he proposed as an explanation for a key feature of the Ice Age: the cyclical periods of glaciation in the "recent" (<1 million year) history of the Earth's climate. The changes he calculated in the tilt of the Earth's axis and the shape of its orbit were incredibly small. Such slight perturbations would only be capable of causing dramatic shifts in the Earth's climate if the planetary weather system was intrinsically "metastable"—capable of slipping into either a much colder or a much warmer condition. Increasingly, climate scientists came to believe this was possible. As a physical mechanism for how this might happen, they proposed the existence of so-called "positive feedback" which could cause temperatures to accelerate. For example, as snow and ice melt, they absorb much more incoming sunlight, further accelerating the rate of melting. Indeed, many such feedback processes have found their way into the modern climate simulation tools that are used today. Scientists began to accept the idea that rapid changes in the Earth's climate had taken place in the past and were possible in the future.

First Evidence of Human Impacts

The next major advance in addressing the possibility of testing the Arrhenius-Callendar global warming theory came through the collaboration of a number of dedicated scientists, led by Dave Keeling, Roger Revelle, and Hans Suess, who persisted through the untimely interventions of both the Second World War and the Korean War to eventually piece together enough government research funding to collect the first accurate data on atmospheric carbon dioxide levels in Antarctica and at Mauna Loa, Hawaii (both far enough removed from local carbon dioxide sources to collect globally-representative information). As shown in Figure 2.1, the Mauna Loa data tell a compelling story about the rate that atmospheric carbon dioxide continues to climb. The annual fluctuations are caused by Northern Hemisphere plants, which consume carbon dioxide during summer months and then release it during the winter months of decay.

Another key technology advance came at about the same time from Cesare Emiliani, a geology student from Italy working at the University of Chicago, who worked out many of the experimental details in a new isotopic method for inferring prehistoric temperatures, based on the presence of a rare nuclear isotope of oxygen, ^{18}O (the most prevalent oxygen isotope has a mass of sixteen). In 1947, the nuclear chemist Harold Urey had discovered that the ratio of ^{18}O in the shells of a class of marine organisms (*foraminifera*) was directly related to the temperature of the water at the time that the organism had lived. Since these shells can be found at the bottom of the ocean in discrete layers that may be simply counted and dated, the past temperature of the Earth's oceans could be directly determined. Once the technical details were worked out, climate scientists had a much better record of historical temperatures with which to test various models of the ice ages and of the climate's true sensitivity to the small variations in sunlight suggested by Milankovitch's calculations.

A series of debates then ensued within the climate science community on whether the Earth's climate was really as sensitive to small perturbations as the increasing amount of experimental data were suggesting. It took time for this to become a widely accepted idea, as it takes scientists (who are, after all, just people) far outside of their "comfort zones." The natural response to examining the vast scale of Creation is to look in awe and be overwhelmed by its age, apparently infinite resilience, and resistance to change (remember the commonly accepted principle of "stationarity"). However, this is an illusion perpetrated by the true infinitesimal nature of our personal experience. Things may seem to be calm and unchanging (except perhaps, in the midst of a severe thunderstorm), but this is not so. Much of the world really was covered by glaciers not so long ago, probably during the time our first ancestors walked the planet. The world really was inundated by sudden floods when the natural dams holding back those glacial lakes suddenly burst. But it took several decades before the implications of these realities sunk into the larger science community.

COMPUTERS AND WEATHER PREDICTION

In order to proceed with detailed climate science of the kind now possible, it was necessary to first develop yet another enabling technology, computers. In 1922 Lewis Fry Richardson had proposed a complete numerical system for weather prediction, in which the area of interest was divided into cells, each with a set of numerical values representing the key weather variables: temperature, air pressure, wind speed, relative humidity, etc. As any good engineering student knows, one can then use the Navier-Stokes equations, a logical consequence of Newton's laws of conservation of mass, heat, and momentum, to directly calculate future behavior of the system.

But there's one problem—well actually, lots of problems. First of all, you don't actually have accurate information to start the calculations. Plus, you end up needing to make lots of simplifying assumptions to do the mathematics, which is a set of coupled, nonlinear differential equations—don't try this at home! The breakthrough came when John von Neumann, a brilliant and ambitious Princeton mathematician, advocated the use of computers for performing the calculations. He had become aware of computers, which had actually been a closely-held military secret during World War II, due to their use in deciphering German and Japanese codes, not to mention the Manhattan Project (to build the first atomic weapons).

Von Neumann's proposal received support from several government agencies and modern weather forecasting became a reality…well, sort of. It actually took at least ten years worth of advances in both the mathematical approximations and the inner guts of computers (still filled with vacuum tubes, like the kind in old TV's) before they were able to catch up with the map-reading human forecasters of the day. Even today, of course, we are all painfully aware of the notorious unreliability of weather forecasting.

It turns out that the inability to predict the weather is not because our computers are not fast enough, or that we have the math all wrong, or even because the Chamber of Commerce bribes the TV forecaster! It has to do with those pesky nonlinear differential equations that govern the system. It turns out that they have an unavoidable technical feature, first explained in detail in 1961 at the Massachusetts Institute of Technology by Edward Lorenz (the Butterfly Effect mentioned in Chapter 1). He had developed a computer program for simulating weather patterns, essentially realizing the idea first dreamed up by Richardson about forty years earlier. One day he decided to repeat a calculation he had previously made, this time for a longer time period. His computer used six significant figures for each of the variables, but he had only printed things out with three digits, so it was these slightly truncated values that Lorenz entered into the program as a starting point for the repeat calculations. To his surprise, the result completely diverged from the original computation within about thirty days of the simulation, resulting in completely different weather predictions!

For the next two years, Lorenz pored over the mathematics behind these phenomena and in 1963 he published his findings, asserting that all such solutions to the governing equations were unstable with respect to infinitesimal changes in the input conditions, thereby rendering all long-term weather forecasting completely impossible, on theoretical grounds. Thus, after some four thousand years, we have an answer to the question first posed by Elihu, "Can anyone understand the spreading of the clouds?" (Job 36:29). Answer: no, at least no mere mortal human being can. This is one of many areas of knowledge that is strictly "off-limits" to man.

Lorenz presented his findings at a 1965 conference on "Causes of Climate Change" in Boulder, Colorado. His colleagues latched onto the result as new evidence for the possibility that climate could quickly "flip" to a new state, including even the oceanic circulation systems, which were then beginning to become known as critical in determining global weather patterns. Another presenter at the same Boulder conference, Peter Weyl of Oregon State University, presented a complicated theory of sea ice formation and its interaction with the salinity (total salt content) of seawater. He specifically pointed out that a drop of salinity could prevent the cold water from being dense enough to sink, thereby stalling the thermohaline circulation pattern in the North Atlantic, and causing a sudden cooling in the Northern Hemisphere. Taken as a whole, the 1965 conference caused wide acceptance of the concept that the Earth's climate was in a precariously balanced state completely susceptible to either rapid heating or cooling in response to even the smallest of nudges.

THE ICE AGE COMETH?

From the mid-1960s into the early 1970s, climate science became engrossed in unraveling a new puzzle that has ended up hurting its credibility in the eyes of the public and has also made it easier for skeptics to poke apparent holes in the current chorus of global warming warnings. The key question was this: was there a danger that man-made pollution could cause drastic cooling due to the continued release of aerosols, particulate matter, and even contrails produced by jet travel? The question received additional attention when researchers found compelling evidence that the Earth was somewhat "overdue" for its next period of heavy glaciation, at least according to the time series of temperature records that were emerging from ice core records (see Figure 2.2). In 1972, these data helped prompt the leading glacial-epoch experts to meet at Brown University and to conclude, "the natural end of the warm epoch is undoubtedly near." There were several naysayers at the conference, but the majority succeeded in issuing a statement saying that serious cooling "must be expected with the next few millennia or even centuries." This press release and the hullabaloo that followed managed to make it to the front page of Time Magazine that year and even prompted a letter of warning to Richard Nixon.

At the time of these cooling warnings, some scientists were instead already concerned with the possibility of global warming from man-made carbon dioxide in the atmosphere, but the majority (so-called "consensus") view at that time was that global cooling was the greater danger, due to the man-made addition of aerosols and particulate matter into the atmosphere (sometimes known as "global dimming"). Looking back at the temperature record for that period now, it seems hard to fault the consensus view. The current spate of warming began in around 1970, and the data for the previous thirty years showed cooling. But this part of the history of climate science shows the intrinsic danger of basing the acceptance of something as true based on a so-called "consensus" of scientists. Many of the world's greatest scientific and spiritual advances have been made by bold individuals (Moses, Jesus, Luther, Newton, Einstein, etc.) who stepped out against the status quo of their day to boldly proclaim truth amidst a chorus of nearly unanimous contradictory views.

Adding to the general cooling fears of the early 1970s was a 1971 paper by S. Ichtique Rasool and Stephen Schneider that predicted a further doubling in atmospheric dust levels might seriously cool the planet by as much as 3.5°C (6.3°F). Schneider, who is now one of the world's leading proponents of the theory of man-made global warming, subsequently discovered key errors in the 1971 paper that are sufficient to completely invalidate the predicted cooling, but at the time the predicted cooling simply added to the ice age hysteria. The entire incident, and especially the press coverage it received, is still a monkey on the back of today's climate science community and may even be causing them to be overly cautious in the way that they describe the peril of our current warming.

Meanwhile, on the other side of the planet, a Russian scientist named Mikhail Budyko had constructed a greatly simplified set of energy balance equations for studying the intrinsic stability of the overall Earth's climate, rather than attempting to predict the idiosyncrasies of the weather patterns. His system had the compelling characteristic that it remained stable so long as the global mean temperature remained in a "normal" range. However, if something caused the temperatures to nudge above or below certain critical points, then his model predicted "runaway" heating or cooling up or down to completely different "newly stable" states. The very warm state looked very much like the conditions believed to have dominated during much of the dinosaur era—far higher temperatures, more carbon dioxide, and much higher sea levels (by hundreds of feet!). At the other end of the thermometer, his model suggested a cool stable state, in which Earth could become a gleaming ball of ice. Though these predictions may seem impossibly preposterous, they were given more credibility when an American scientist, William Sellers at the University of Arkansas, derived similar results, albeit it with a very different mathematical model. Both models suggested that the Earth's climate had natural instabilities that could cause accelerated warming in response to seemingly small perturbations, such as the increases in the trace atmospheric concentrations of carbon dioxide and other greenhouse gases that were then being measured accurately for the first time. The possibility of man-made global warming was now becoming more real in the minds of leading climate scientists.

Hell on Earth?

Looking to the heavens, some climate scientists and astronomers (including even Carl Sagan!) began to develop models to explain how Venus became the hellish hothouse that it is. It too experiences a greenhouse warming effect analogous to Earth's. But the effect is not nearly so comfortable. On Venus, the mean temperature is 464°C (867°F), rather than the too cold −46°C (−51°F) that it would be in the absence of the greenhouse effect. A scientist named Andrew Ingersoll published a 1969 paper showing that it could have reached this state after first beginning with a much more Earth-like climate dominated by water, all of which has now long since evaporated. Venus's current atmosphere is a thick layer of mainly carbon dioxide overlaid by clouds of sulfuric acid, which form *real* acid rain pouring down in a manner strangely and eerily reminiscent of the fire and brimstone mentioned in Scripture. Ingersoll showed that Earth would have suffered a similar fate had it been only 6% closer to the sun. Fortunately for us, God had other plans! He placed the Earth just far enough away that the negative feedback of increased cloud cover caused the climates of the age of dinosaurs to be very warm, but not so hot that at least some forms of life could not persist. Our current warming trajectory appears headed toward that same sweltering climate (see Figure 2.3). However, it appears it will take only a few more decades, as opposed to the approximately sixty-five million years (according to modern climate science), required to cool to the temperate climate within which all of civilization developed.

During the 1970s, the question of rising carbon dioxide levels in the Earth's own current atmosphere would occasionally still come up, based largely on the vocal advocacy of Roger Revelle, Stephen Schneider, and Reid Bryson. But it received little traction in either the larger scientific community or the public, since there was still no convincing evidence of a global warming trend at that time. Schneider and a colleague published an apparently prophetic paper during this period, suggesting that warming due to higher carbon dioxide levels would soon begin to dominate the Earth's climate after 1980. A 1977 National Academy of Sciences panel issued a report also suggesting that catastrophic warming, not cooling, was the greatest threat to the Earth's climate. But this all came too soon after the 1972 Brown University group's warnings of an imminent ice age to win very many converts. At the end of the decade a World Climate Conference was held in Geneva in 1979, convening 300 experts from 50 countries. They issued a consensus statement recognizing the "clear possibility" that an increase in carbon dioxide "may result in significant and possibly major long-term changes of the global-scale climate." This was hardly news-grabbing language, but it was enough to keep some momentum going for additional research.

The 1980s saw the development of the first true Global Climate Models (GCMs) by independent teams of researchers from around the world. These models differed in many fundamental ways from both weather forecasting simulation models and the very simple energy balance approaches that had been proposed years earlier by Budyko and Sellers. Among the key differences were the addition of a true oceanic circulation model, some representation of land topography, and several feedback processes. The last issue is absolutely essential to the models and warrants a brief diversion from the history discussion.

Feedback: The Key to Accurately Predicting the Rate of Warming

Just as in personal relationships, it turns out that feedback is a key factor when it comes to the theory of man-made global warming. But climate scientists use the term in a particular way and define two types of feedback: positive and negative. Positive feedback processes have the potential to greatly accelerate the rate of warming. A simple example is the melting of snow or ice, both of which reflect sunlight. After they melt, they typically expose darker surfaces (soil or open water), which absorb more incoming solar radiation and therefore increase the rate of warming. On the other hand, negative feedback would tend to retard warming and act more like a thermostat to keep temperatures where they are. A simple example of this is cloud formation. As the ocean warms, more water evaporates, but this increased atmospheric water content could increase cloud cover, which would tend to reflect more sunlight back out to space, thereby cooling the Earth.

An obvious question is how to conduct a scientific experiment to convincingly prove whether these feedback processes actually belong in the model, and how strong they really are. Most of the processes (melting of ice, plant response to higher carbon dioxide, decomposition of organic matter in the permafrost, etc.) are so complex and operate at such a vast scale that they are not subject to the conventional scientific method of experimental design (with proper controls and the like). So the main approach has been to build them into the computer models and then use the degree of agreement with the observed temperature trends to see how important they are. But this process has many intrinsic uncertainties due to questions about the reliability of the temperature record and the high degree of empiricism that is necessarily involved in attempting to simulate so many highly complex physical processes (such as cloud formation and precipitation patterns). Another challenge is to do all of this in a world that is quickly becoming warmer than at anytime during the recent period of truly reliable global observations (since the late 1800s).

I see a real potential for the consensus-driven, politically-influenced nature of the model development approach being sponsored by the United Nations to end up with climate models that are too conservative and underestimate the strength of positive feedback processes that govern the rate of global warming. It seems unlikely there will be a way to escape this "human element" of the climate modeling process.

But I digress. Let's get back to finishing off the history discussion.

THE TIPPING POINT

By the mid-1980s political pressure began to grow, first in Europe and eventually in the United States, for "something to be done" about the global warming issue. Although he failed to win the nomination, Al Gore was a leading presidential candidate on the Democratic side, and he made concerns over global warming one of his key issues during his 1988 campaign. The first "tipping point" came during that hot summer of 1988, when much of the Midwestern United States was suffering a prolonged drought and an unusually hot summer. Responding to all of these pressures, the United States finally relinquished its veto power and the United Nations created the Intergovernmental Panel on Climate Change (IPCC), which now continues to lead the world efforts in this area, with a considerable amount of funding and political clout. Many of today's leading climate scientists chose to join the IPCC, which has since issued a series of four detailed assessment reports: in 1992, 1996, 2001, and 2007.

By the late 1980s, the science was also getting helped by the wider availability of satellite-based imagery for measuring temperatures and global cloud cover. Better ice-coring methods were also developed and implemented for measuring the long term temperature record and the time series of atmospheric concentrations of the major greenhouse gas constituents: carbon dioxide, methane, and nitrous oxide.

The first IPCC communication of any kind came in 1990, with the very cautious, consensus-driven language that has characterized all of its subsequent statements. The statement concluded that there had been warming but that much of this may have been due to natural processes with very little evidence of a dominating man-made component. Given the apparently valid predictions of Schneider from just over a decade earlier, which stated that carbon dioxide driven warming would only become dominant after 1980, this hardly seems surprising. But the 1990 report generated very little news, coming as it did in the happy afterglow of the Berlin Wall's recent fall and the beginnings of the first Gulf War.

But the 1990 report did attract attention from some, particularly those on the conservative side of the political spectrum who were uncomfortable with the United Nations, the US Government, and anybody else that stood in the way of unbridled, free market utilization of natural resources such as oil and coal. Although these political and industry interests did not form many visible alliances, there began to be a groundswell of "contrary" views harshly denigrating the efforts and motives of the IPCC. To be sure, many of these criticisms came (and still come to this day) from the considerable number of sincere scientists who have been unconvinced by the evidence presented by the IPCC. A number of different arguments have been raised, but the overall theme has been to cast considerable doubt on the theory of man-made global warming.

One of the leading skeptics who emerged to challenge the IPCC dogma was S. Fred Singer, who retired from a distinguished government service career to found the Environmental Policy Project, with funding from generally "conservative" sources. Singer and others began to mine the IPCC reports for nuggets of evidence that were damaging to the credibility of the models. For instance, the first set of modeling runs predicted a temperature rise roughly double what was actually believed to have taken place. There were also endless questions about the reliability of the temperature record, due to the "urban heat island" effect and other similar problems with the underlying data. Throughout most of the United States, Gore's continued warnings about global warming became somewhat of a running joke among Republicans and their supporters. And the global warming advocates always had to be prepared to be reminded of the ice age scare that they had helped perpetrate in the early 1970s.

The media in the United States were going through their own cable news to Internet transformation, in which they became compelled to broadcast ever more shrill "conflict-news-talk" formats. On these programs, it was necessary to portray every issue with dueling adherents who were anxious to appear as combative as possible, which most scientists shied away from. So the theory of man-made global warming took a back seat to such issues as the personal lives of Bill Clinton, Newt Gingrich, and other polarizing politicians. However, newspapers, magazines, Internet news sites, and finally blogs rushed in to fill the void, such that it is now possible to become immediately inundated by a confusing stream of conflicting reports on every aspect of global warming after a few quick keystrokes on Google.

Meanwhile, the IPCC process slogged along with increasingly complex models and an exponentially growing body of peer-reviewed science to back up earlier propositions. The political aspects of the United Nations effort involved a series of widely covered conferences: Montreal (1987), Rio (1992), and Kyoto (1997). The so-called Kyoto Protocol, which was endorsed at the 1997 conference through some intense international negotiations, was adopted by every major nation of the world but for two notorious holdouts: the United States and Australia (however, Australia finally ratified it in December 2007). Kyoto calls for a rollback in greenhouse emissions by the developed countries to 1990 levels by the year 2010 except for India and China. It also exempts big sources like airline travel and is a very imperfect and largely symbolic move at best. The "Bali Roadmap," just recently negotiated as this book has gone to press, is not an actual emission-reduction plan but is instead the deferral of any real negotiations until a new deadline of year-end 2009. The cynic in me expects the next IPCC plan to be just as ineffective as the earlier Kyoto Protocol. But only time will tell.

The overall effect on the American public has evolved to our present situation, which seems to be generally wearisome of hearing any more about the shrill scientific debate and scary warnings. Skeptical scientists have continued to lob volleys against the prevailing theories. They have argued, for example, that variations in solar activity could explain all of the recent warming. Richard Lindzen of the Massachusetts Institute of Technology challenged the way that the modelers had simulated water vapor feedback. The skeptics also objected to the models because they had still failed to correctly predict any actual time series of measured historical data. But there also seems to be a general willingness to accept varying degrees of government action—so long as it doesn't come at the expense of actually having to restrict our consumptive lifestyles.

So what does the IPCC actually have to say about what's happening with our global climate right now?

IPCC Assessment of the Current Situation

The most recent IPCC report (2007) reflects considerable progress based on large amounts of new and much more comprehensive data, improvements in the understanding of the underlying processes, and more sophisticated analyses of the model results. All of these factors enable better characterization of the uncertainties in the climate predictions. The report quantifies the relative impacts of man-made and natural factors in terms of "net radiative forcing" in units of energy per unit area (watts per square meter). According to the IPCC, the most important factors include changes in the abundance of greenhouse gases, particularly carbon dioxide, methane, nitrous oxide, and chlorofluorocarbons (CFCs). They conclude that the changes brought on by the increasing concentrations of these gases have a significantly greater effect than the other factors, such as man-made ozone, albedo (surface reflectivity) effects, aerosols (direct and indirect via cloud formation), and variations in solar activity. Of all the other factors affecting climate, the IPCC scientists currently believe that the largest cooling factor is the presence of man-made aerosols in the atmosphere, which are just enough to offset all of the warming factors except for carbon dioxide, which ends up driving the overall global system in the direction of warming.

The amount of new radiative forcing directly attributable to man-made greenhouse gases (nearly 3 watts per square meter) already equates to about a 2% increase in absorbed sunlight, which is like moving the Earth a million miles closer to the Sun! This dwarfs natural variation in solar intensity, which is on the order of 0.1%. It also dwarfs the direct warming effects of orbital wobbling calculated by Milankovitch, which is only about 0.5 watts per square meter, yet sufficient to cause the global temperature swings of 10°F that have been causing repeated cycles of glaciation over the past hundreds of thousands of years.

Global mean carbon dioxide levels in the atmosphere have increased from a preindustrial (1750) level of about 280 parts per million (PPM) to

379 PPM in 2005. This current value far exceeds the highest level in the ice core record of the past 650,000 years, which was 300 PPM. The current annual growth rate in atmospheric carbon dioxide is around 1.9 PPM per year, which is a considerably faster rate of growth than what was observed (about 1 PPM per year) when Keeling and his colleagues first began measuring atmospheric carbon dioxide in the late 1950s. The primary source of the increased atmospheric carbon dioxide is fossil fuel use, with land use changes (destruction of forests and other natural systems) providing a smaller contribution.

In order to discuss global man-made emissions of carbon dioxide, it will be necessary to introduce the uncommonly heard term, gigaton, which is directly analogous to the computer term, gigabyte. In both cases, the prefix "giga" means one billion. In this case the suffix, "ton," is not the English ton of American parlance (2,000 pounds) but actually refers to a metric tonne (1,000 kg). Suffice it to say, a gigaton is a "whole bunch" of stuff. Fossil fuel based emissions of carbon dioxide increased from about 23.5 gigatons per year in the 1990s to 26.4 gigatons per year in the 2000–2005 period, whereas the contribution due to land use changes during the 1990s was less, in the range of 1.8 to 9.9 gigatons. The atmospheric carbon dioxide caused by land use is subject to greater relative uncertainty, but it is clearly much smaller than the amount delivered by the combustion of fossil fuels.

As for the concentrations of methane, it has increased from a preindustrial (1750) level of 715 parts per billion (PPB) to 1774 PPB during the industrial age, again far higher than the highest known level during the previous 650,000 year period: 790 PPB. This gas is believed to be produced primarily as a result of agricultural activities (livestock), but there are other possible sources, such as anaerobic decomposition of the permafrost. The other important greenhouse gas derived from agriculture is nitrous oxide, which has increased from 270 PPB to 319 PPB during the industrial period.

When combined with the cooling effects of aerosols, the IPCC concludes that the net anthropogenic (man-made) effect on the climate is more than 90% likely to be one of considerable warming, with an overall magnitude roughly equal to that of the warming influence of carbon dioxide alone. The current rate of warming is more than 90% likely to be unprecedented in more than 10,000 years. According to the IPCC, the rate of warming is accelerating. They quote a current global mean temperature rise of 0.13°C (0.23°F) per decade for the last 50 years. As discussed later in this book, Northern Hemisphere land surface temperatures are increasing at an even faster clip, with significant acceleration as well.

One of the frequent criticisms of IPCC global warming analyses is that the temperature record itself is tainted by the "urban heat island" effect. The idea is that many of the stations recording long-term temperature trends are often positioned at locations where urbanization has produced a more extreme rate of warming than what is truly representative of the entire landscape. Numerous allegedly accurate photographs of particularly extreme cases such as thermometers positioned immediately next to air conditioner units and the like are now proliferating on the Internet—perhaps causing Mr. Gore to now regret having "invented" it! The IPCC addresses the urban heat island effect by pointing out that it is real but has a negligible influence of less than 0.006°C per decade over land and zero over the oceans. The 2007 report states that balloon-borne and satellite measurements of atmospheric temperatures now show warming rates comparable to the surface observations, reconciling an apparent discrepancy from the 2001 report.

The IPCC also reports that global mean atmospheric water content has increased since the 1980s over land and ocean and throughout the atmosphere, a fact that is in broad agreement with the models and with the higher global mean temperature. This is perfectly logical because more water will evaporate from the warmer surface waters of the ocean, which also implies a global mean increase in precipitation. The observation also belies a complex feedback mechanism, because water vapor is the most important greenhouse gas in terms of its global capacity to absorb infrared radiation. However, condensed water vapor (clouds) reflects sunlight away from the lower layers of the atmosphere, a negative feedback process of unknown strength and impact.

As for the oceans, IPCC states that temperature records of ocean depths now show a warming signal propagating down to depths of at least 3,000 m, with the oceans absorbing more than 80% of the additional heat energy that has been pumped into the Earth as a result of global warming. The ocean's higher temperature causes it to expand, which has contributed to sea level rise. Also contributing to the observed sea level rise is the melting of glaciers and polar ice caps. The total sea level rise during the twentieth century was 0.17 m (6.7 inches), with the rate of rise increasing to a rate of 31 mm (1¼ inches) per decade at the end of the twentieth century.

The 2007 IPCC report is the first from the panel to discuss a very troubling and recently discovered man-made impact on the sea: ocean acidification. New data show that at least half the carbon dioxide produced by man has been absorbed by the oceans, and this has already measurably changed its degree of acidity. Anyone who has taken high school chemistry should remember the pH scale, which is a direct measure of acidity. Lower numbers on the pH scale (less than seven) are indicative of acids (would taste sour, like vinegar); whereas higher numbers (more than seven) are for solutions dominated by the presence of bases (would taste bitter, like unsweetened chocolate). The point of this is that the continuing uptake of excess carbon dioxide by the oceans has already lowered the pH of surface seawater by 0.1 pH units. This might not sound like much, but as the pH scale is logarithmic, this equates to a 30% increase in hydrogen ion concentrations.

This is a new observation and its implications have not been fully examined, but one very disturbing consequence of this acidification has already been documented. As the pH drops and acidification continues, the solubility of calcium carbonate, the chemical that forms the shells of many marine organisms, will increase. The species at risk include coral, mollusks, and a number of microscopic organisms. As the shells of these marine organisms dissolve they will become increasingly stressed and would be unlikely to survive under even more acidified conditions, potentially threatening entire marine ecosystems. The Royal Society, a British science organization known for using very carefully chosen words, was one of the first to sound loud alarm bells about this newly discovered phenomenon. In a recent report (2005), they concluded with a plea to reduce the growing amount of carbon dioxide that is being absorbed by the oceans, "no option that can make a significant contribution should be dismissed."

Getting back to the IPCC status report of the overall current situation, they cite numerous changes that have become apparent as we begin the twenty-first century. The Arctic is warming more rapidly than other parts of the planet, its amount of sea ice declining by 2.7% per decade over the past thirty years. The area covered by permafrost is also decreasing. Higher precipitation is now being seen in mid to high latitudes, but areas of drought are developing in the Mediterranean, southern Asia, and much of Africa. More intense and longer droughts have been observed over wider areas since the 1970s, especially in the lower latitudes. The frequency of heavy precipitation events is increasing, with a higher proportion of rainfall occurring in the form of thunderstorms. Throughout the world, the number of days with frost is decreasing, and heat waves have become more intense and more common. Although the data are not entirely consistent on tropical cyclones, there is a trend for an increase in the number of intense storms, which appears to correlate with higher sea surface temperatures.

The IPCC states that the warming that has already taken place due to greenhouse gases would have likely been greater had volcanic eruptions and man-made aerosol pollution not intervened to partially offset the amount of radiative forcing. Their final conclusion about the current situation is that it is extremely unlikely that the widespread warming and ice melting now being observed can be explained without some form of external forcing with no known natural causes capable of producing these changes. Having given the august IPCC panel the attention it is due, I'd like now to address the criticisms and comments of the leading climate change skeptics.

Temperature Record Controversies: The "Hockey Stick Graph"

As noted previously, a key point of criticism continues to be whether the temperature record is truly reliable. The debate "heated up" significantly with the publication, in 1999, of the now infamous "Hockey Stick" graph by a series of authors led by M. Mann. The graph has now earned its own separate entry in the Wikipedia online dictionary and has been the object of a seemingly relentless onslaught of criticism. This all goes to show that it must be telling a story that many feel very uncomfortable about—temperatures are rising very rapidly now when compared with any period over the past 1,000 years or so.

Although there is still considerable online debate about the reliability of the temperature record and the Mann paper in particular, most of the underlying data appear to be sound and have been extensively vetted through the scientific peer review process. Within the United States, the key source of these data is the US National Climatic Data Center in Asheville, North Carolina. In a series of recent publications (see Appendix 1), researchers from these government agencies present an improved reconstruction of the sea surface temperatures for the past one and a half centuries. Uncertainties in the record are greatest during the nineteenth century (0.4°C) and during the two world wars (0.2°C), but uncertainties are typically on the order of only 0.1°C for the last half of the twentieth century, during which the authors report there has been significant warming, especially since 1970.

These same authors subsequently built upon their previous sea surface temperature work to add the land surface temperatures and thereby construct global mean temperature records. They provide uncertainty measures that account for both sampling and systematic bias errors and note that the data may now be sampled at finer spatial scales if that is desired. The data extend from 1880 to present, and the authors report that anomalies up until about 1970 all appear to be within the experimental error, but the trend since that time is statistically significant in the direction of global warming.

My personal position on the temperature record controversy is partly informed by a simple practical issue. If we can't get accurate temperature data from the US National Weather Service, then I have no idea where else on the planet we would go in order to get reliable information. In addition, I find the other observations from around the planet to be very convincing: receding glaciers, disappearing Arctic ice cap, rising sea levels, changing animal migration dates, etc. To claim otherwise seems silly! But let's move on to the critics of the theory.

Today's Leading Skeptics

Of course the theory of man-made global warming is just that...a theory, and it has attracted a series of very vocal and sometimes very well funded detractors. Some of the leading opponents are highlighted here. To those I have missed, keep trying!

S. Fred Singer

Already mentioned earlier in this chapter, Fred Singer is one of today's leading skeptics, and he is certainly well qualified. He is currently Professor Emeritus of Environmental Sciences at the University of Virginia, a Distinguished Research Professor at George Mason University, and president of the Science and Environmental Policy Project, an organization that seems to exist primarily to challenge the science and policy recommendations of the IPCC. Singer performed his undergraduate studies at Ohio State University and earned his Ph.D. in Physics from Princeton University. He was the founding dean of the School of Environmental and Planetary Sciences at the University of Miami, the founding director of the US National Weather Satellite Service, and served for five years as vice chairman of the US National Advisory Committee on Oceans and Atmosphere. He has written or edited over a dozen books and monographs, including, most recently, *Unstoppable Global Warming: Every 1,500 Years*.

In a recent article attacking the theory of man-made global warming and intended for lay (not scientific) readers, Singer states that, "the climate has not warmed in the past eight years, even though greenhouse gas levels have increased rapidly" (2007). This is not Singer's only argument against the theory, but it is a particularly weak one—and he is not being scientifically honest when he makes it to his lay readers. Given his extensive background and knowledge of climatic data, he is certainly aware of the very noisy character of the global mean temperature record (see Figure 2.4). By choosing one particular eight year comparison (Singer doesn't say which specific two years he was comparing, but 1998 and 2006 are probably the pair he selected), he is guilty of what is known in the scientific world as "cherry-picking" the data. He has blatantly selected a single contrary observation out of a large and noisy set of data that—when examined in its entirety—completely refutes the point he is attempting to prove!

Some of the other points made by Singer in the same article are listed below in italics, and I have provided a brief commentary on each of them.

- *The geologic record shows a persistent 1,500 year cycle of warming and cooling extending back at least one million years.*

If there is such a cycle, it doesn't seem to be very dramatic in magnitude when looking at Figure 2.2. So I don't see how this even speaks to the question of man-made global warming. At least it doesn't seem relevant to the type of warming already underway.

- *The so-called "scientific consensus" of 2,500 IPCC scientists is an illusion – skepticism among climate scientists is well over 50%.*

I agree with Singer that the terminology, "scientific consensus", which is often used by the IPCC and other advocates for the theory of man-made global warming, is not useful or even relevant about whether a scientific theory is true. Most scientists are unwilling to take a solid position on an issue of such profound significance. My own experience is that they typically have personality types that cause them to leave the room if voices rise and a debate begins to flare up into an argument. But my read of the literature and recent polling data would suggest that of those scientists willing to express a definite opinion, the majority believe the theory is true. It's just that most prefer to sit on the fence.

- *Even if there were such a "consensus," that would not be proof (think of Galileo or Einstein).*

Again, I agree with Singer that "consensus" is not how science works. Of course, neither is cherry-picking the data, something of which Singer is apparently guilty in adopting his very shrill advocacy position. Science is a search for truth, in which one objectively evaluates all of the information—with neither bias toward the majority opinion nor motivated by a personal agenda.

- *The observed pattern of warming, such as the rate of warming at high atmospheric elevation near the tropics does not match the current computer models.*

Given the complexity of the phenomena being modeled, I do not find it surprising that the models have difficulty replicating all characteristics of the global warming process. I'd be more suspicious if the modelers claimed perfection. The honest ones don't.

- *Current models predict faster global warming than actually observed, and are therefore missing an important negative feedback process, such as water vapor induced cloud formation and subsequent higher reflectance of sunlight.*

I completely disagree with the premise of this statement. As I will discuss in Chapter 4, it is my observation that the current models are too sluggish in their predictions for the rate of temperature rise, and that the models are therefore either missing an important *positive* feedback process, or are making some other error that masks the actual cumulative warming effect of the greenhouse gases.

- *Variations in solar radiation are likely the most important factor in current warming.*

This subject became an intense issue for scientific investigation during recent years, and the most recent IPCC report clearly states that the magnitude of this effect is tiny compared with greenhouse gas forcing and other effects, such as the cooling impact of certain air pollutants.

Singer also goes on to question the non–fossil fuel based energy investments now being advocated as all being counterproductive and needlessly expensive. He also questions the wisdom of "emission trading" as simply being a way for corrupt officials to benefit financially from rigged systems. In so doing he alludes to the Iraq "oil for food" scandal that recently plagued the United Nations. Singer concludes that a little warming would actually be beneficial to many world areas, especially the currently frigid regions of Canada and Siberia. Finally, just like Michael Crichton, the next skeptic to be discussed, Singer calls out the immense financial interests of environmental advocacy organizations that further the "hysteria" via the media and thereby induce additional donations—as a "positive feedback process" not unlike those that are believed to help fuel global warming in the climate models themselves.

MICHAEL CRICHTON

Crichton is a hugely successful and influential novelist entrepreneur, having created the long-running and hard-hitting television series, *ER*, and the wildly popular movie series, *Jurassic Park*, among numerous other impressive works. In the course of writing his 2004 novel on man-made global warming, *State of Fear*, Crichton states that he did about three years of research, which caused him to develop a deeply-held skepticism about the theory and its proponents. In a postscript to the 2004 novel, Crichton sheds the role of novelist and plainly states his personal feelings about the theory, labeling them as "Author's Message." He concludes that we really can't know how much, if any, of the current warming is man-made and launches into various attacks against those who exploit this uncertainty to their own advantage, financial or otherwise. Although Crichton is right that nothing can be known scientifically with absolute (100%) certainty, it is possible to make meaningful statements with known levels of certainty (e.g. >90%), to which we may appropriately attach adjectives such as "very likely."

He also describes today's global warming theorists as having been deluded by "ignorance" or "pigheadedness" in warning about the unsustainability of today's global development trends. On the contrary, it seems less rational to me to believe that a finite planet with only one external energy source can support unending exponential growth in energy consumption and pollution generation while at the same time sustaining the narrow balance of environmental conditions necessary for life as we know it. The basic laws of thermodynamics will prevail. At times, Crichton seems to vastly overestimate the ability of mankind to fake our way out of this predicament—at one point painting a utopian picture of life on Earth in the year 2100. Elsewhere, he states that humanity always messes things up when it attempts to intentionally restore natural ecological systems. His latter assessment is likely closer to reality.

Conservative Evangelicals in the United States

Most conservative evangelical organizations in the United States have taken a very dim view of the entire theory of man-made global warming. For the most part, I believe this skepticism has been prompted by the same factors that motivated my own original opinions—a "knee-jerk" negative response to the people (Al Gore) or organizations (the United Nations) most associated with advocating the theory. The Center for Reclaiming America for Christ is one such group. It is one of the many successful and influential ministries founded and led by Dr. D. James Kennedy, who passed away in mid-2007.

Prompted mainly by Al Gore's 2007 Oscar Award for *An Inconvenient Truth*, one recent publication (2007) from the Center for Reclaiming America for Christ adopts a very negative view about the theory. For the science around this subject, the article relies mainly on the comments of Dr. E. Cal Beisner, Associate Professor of Historical and Societal Ethics at Knox Theological Seminary. While acknowledging the fact of some recent warming, the article says this might not be a bad thing and that there is little evidence that humans are to blame. For additional support of this latter point, the article quotes a familiar and previously mentioned climate change critic, Richard S. Lindzen, professor of meteorology at the Massachusetts Institute of Technology (MIT), "all that can be asked is whether the observed changes in climate are so large they can be distinguished from the regular changes the Earth undergoes. For the moment they are not." The article concludes with a familiar theme among conservative evangelicals, which is that increased government spending to curb carbon dioxide emissions will hurt the poor the most, by driving up prices for fuel and food, for instance.

A short sidebar article in the same publication describes a controversy stirred in late 2006 and early 2007 when Rick Warren (of *Purpose-Driven Life* fame) and eighty-five other evangelical leaders issued a call to action for Christians to curb greenhouse gas emissions via a print and TV ad campaign in *The New York Times*, on Fox News, and elsewhere. Among the evangelicals who declined to endorse the 2006 initiative were Dr. James Dobson, Charles Colson, and Dr. D. James Kennedy. In early 2007 Dobson led an effort to have a National Association of Evangelicals (NAE) official,

Richard Cizik, fired for "dividing and demoralizing the NAE" by "using the global warming controversy to shift the emphasis away from the great moral issues of our time, notably the sanctity of human life, the integrity of marriage, and the teaching of sexual abstinence and morality to our children." This entire episode provides ample motivation for this book, as it reveals a deep rift among evangelicals over the issue of man-made global warming. And we know who has the most to gain from creating such a rift: the enemy of our souls. It is my humble prayer that this book will play a helpful role in shoring up this rift and getting my brothers and sisters in Christ focused on the most important issue—winning souls for Christ!

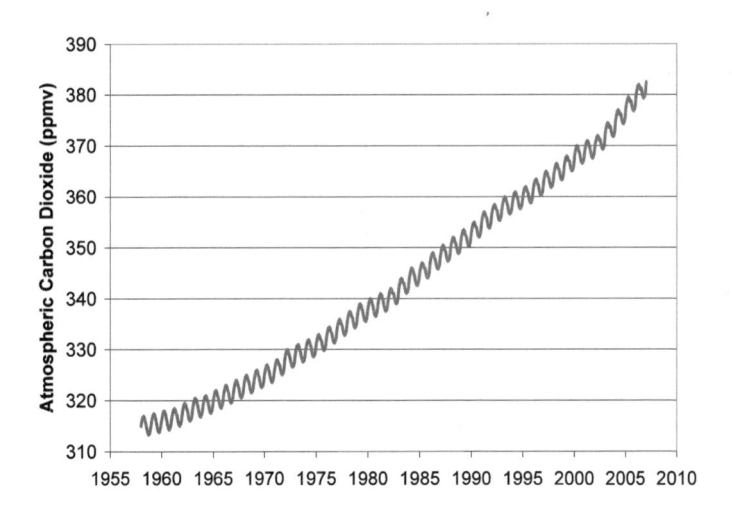

FIGURE 2.1 Carbon dioxide levels in the atmosphere, as measured at Mauna Loa (source: Keeling et al.).

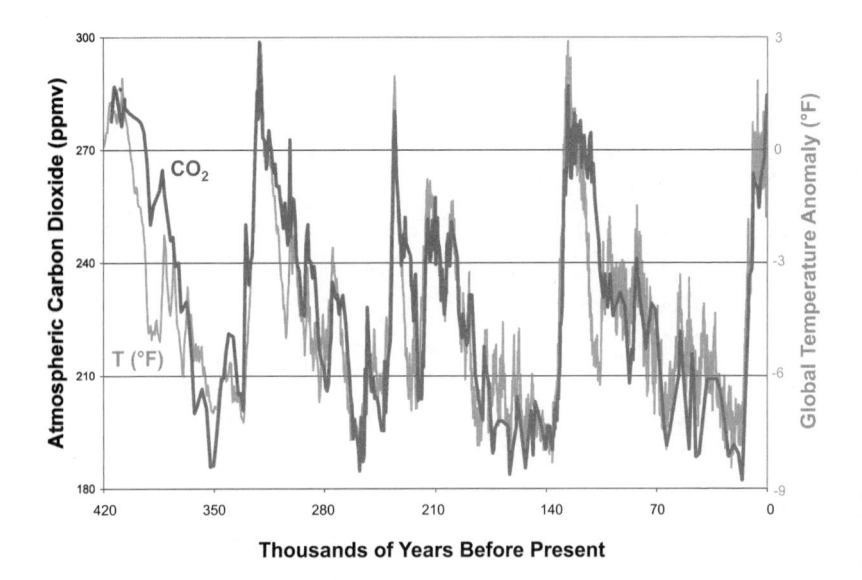

FIGURE 2.2 Global mean temperatures and CO_2 levels for the past half million years, based on the analysis of air bubbles trapped in ice cores collected on Antarctica (source: Vostok ice cores).

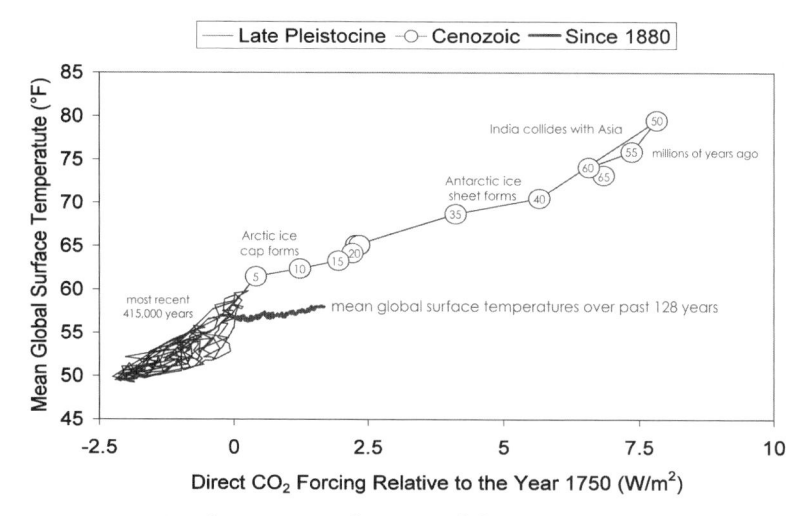

FIGURE 2.3 "Phase space" diagram of the Earth's climate for the past sixty-five million years that shows the unprecedented speed of the warming trajectory we are now following (replot of data from NOAA, Vostok ice cores, and Hansen et al.).

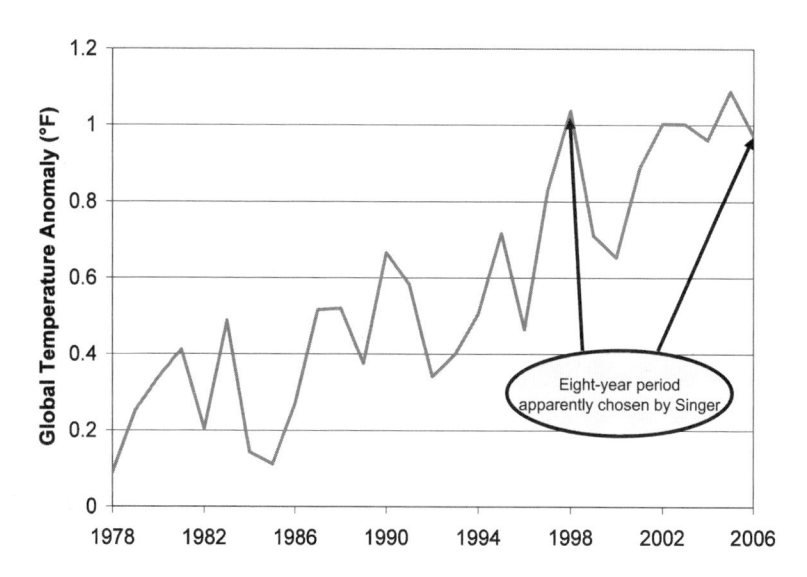

Figure 2.4 Global mean temperatures for the past thirty years, which contradicts the intentionally misleading selection of contrary data by S. Fred Singer (source: NOAA).

Chapter Three

Climate Change in the Bible

As we have just learned, modern science teaches that the Earth's climate has been constantly changing with dramatic and rapid changes possible at any time. The Bible also speaks of climate change—some of it (such as the Flood of Noah's day) even more dramatic and more rapid than anything most modern climatologists would accept as literal. But this is not the only Biblical account of dramatic weather: climate figures prominently in many books of the Bible, especially the first (Genesis) and the last (Revelation). In this chapter I will set out to catalog these references to climate and God's perspective, as revealed in Scripture, concerning man's role in climate change and our part in the ultimate fate of the Earth. As explained in Chapter 1, I will undertake this entire discussion with a view that the Bible is of divine origin and that the truths it contains about both our past and our future will outlast the test of time. In adopting such a position, I am fully aware that this might seem strange and "unscientific" to many a modern reader, but I am not ashamed to confess my faith in the Bible and the triune God who authored it.

But I am also a practicing scientist. And therefore I will occasionally comment upon and perhaps even critically examine the Biblical texts from this scientific perspective. However, as a scientist, I have found through simple empirical evidence that the Bible has withstood every logical attack ever launched against it, just as I have found that it satisfactorily explains the source of the evil within my own heart. However, I am also forever grateful that the text doesn't stop there. In these same Scriptures we find the universal offer of the free gift of eternal life, a life we will continue to enjoy in resurrected bodies long after this Earth has been utterly destroyed. Does that truth mean that we should live our lives today with a callous disregard for the negative consequence our actions have on this planet and toward our fellow inhabitants? Absolutely not! I will answer this question more fully in Chapter 6, but for the moment, let us begin our Biblical journey at the beginning, in the book of Genesis.

In the Beginning ...

Genesis clearly teaches that the Earth's climate has been changing, at least since man first appeared. The first reference to climate is given early in the second chapter of Genesis, "Now no shrub of the field was yet in the earth, and no plant in the field had yet sprouted, for the Lord God had not sent rain upon the earth; and there was no man to cultivate the ground. But a mist used to rise from the earth and water the whole surface of the ground" (Gen. 2:5–6). This detailed description of a climate completely different from our current precipitation patterns is actually consistent with some computer model simulations of conditions that are thought to have prevailed during the most recent period of extensive glaciation (which began a little over 100,000 years ago, see Figure 2.2). This is the same time frame in which uncontested evidence for our species, *Homo sapiens,* first appears in the archaeological record.

How Literally Should We Take the Bible?

At this point, I feel compelled to explain my position on how literally one should take the Bible, especially when it comes to "scientific facts." This is a complicated question to answer, because the Bible contains many literary forms, not all of which are intended to be taken in a completely literal manner. Genesis is a perfect example of this, in which one finds a beautiful tapestry of poetry, history, and spiritual truths, all ending with the story of a principled and faithful hero—Joseph.

As elsewhere in the Bible, Genesis tells the story of Joseph in a way that emphasizes the parallels of his life to the eventual mission of Jesus—who is the central character of all Scripture. Joseph saved the very brothers who cruelly sold him into slavery, and in so doing his story serves as a sort of "coming attractions" movie trailer for the subsequent story of Jesus, who saved all of us, the very ones who mutually sent him to the cruel cross of Calvary. As it builds towards this climactic story of personal integrity and godly wisdom, Genesis contains extensive historical narrative, which I generally regard as factual or literal, although I am quick to admit that many statements stretch the credulity of my corrupted modern mind, especially those regarding the apparent extreme longevity of our ancient ancestors.

Although it is obviously not the main subject of this book, I would simply note that modern science is now discovering simple genetic methods for slowing the aging process dramatically, and it would not surprise me at all for us to eventually learn that the many mutations that now spoil our respective genomes have dramatically shortened our life spans relative to those of the biblical patriarchs.

Throughout its remaining sixty-five books, the Bible contains other forms of literature, including additional historical narratives from the pre-Exilic era (Exodus through Chronicles), songs (the Psalms and the Song of Songs), wisdom literature (Job, Proverbs, Ecclesiastes), personal vignettes (Esther, Ruth), thundering prophecies (Isaiah, Jeremiah, Ezekiel, Daniel and the so-called minor prophets), the four Gospels (Matthew, Mark, Luke, and John), the epistles (letters from Paul, James, Peter, John, and Jude)—and the dramatic final book: John's Revelation (Apocalypse). Across the spectrum of the literary forms contained in these books, there are varying degrees to which they contain the type of refutable declarative statements for which it is legitimate to ask whether the text should be accepted as literally true. There are many thousands of such statements, and a few of them admittedly seem to be "historical hyperbole."

But the overall point of the Bible is not whether every historical statement is a scientific fact in the sense that we use this term today in modern discourse. The eternal spiritual truths about Jesus that are contained in the Bible far exceed the power of a simple chronological listing of facts. Many of the most contentious biblical statements are impossible to independently verify at this time anyhow, so there seems to be limited value in dwelling on this question. As I write this, the human family finds itself teetering on the edge of various global calamities: the environmental ones that are the subject of this book and undeniably "human" ones such as the proliferation of weapons of mass destruction (WMD) and terrorism.

Thus I am far less concerned with questions about the *origin* of our species than I am with questions about our *destiny*. Having said this, I personally believe the Bible contains accurate information about both ends of the timeline of our species and everywhere in between. However, the argument about whether every historical statement across over 2,000 years and over 40 authors should be accepted as a scientific fact entirely misses the main point. The supreme concern of the present hour is whether the Bible's prophetic statements about the catastrophic global events predicted to immediately precede Jesus' return to the planet are true. This leads us to the issue of prophecy.

LITERAL FULFILLMENT OF PROPHECY

As a scientist, I fully recognize that fulfilled prophecy is a direct violation of the "natural laws" that appear to govern this universe. However, my testimony is that God used several personal experiences of such fulfillment in my own life to cause me to recognize that there is a Higher Power outside of Creation who is able to make the river of time flow uphill. I am convinced that God uses the reality of such literal fulfillment as a club over the head of particularly prideful and stubborn cases (such as myself) to cause us to realize the true grandeur of His authority over the universe. But the chief purpose of all biblical prophecy is to reveal the true identity of the person and mission of Jesus. After His resurrection, He pointed this out Himself to the two grieving disciples on the road to Emmaus, "And beginning with Moses and with all the prophets, He explained to them the things concerning Himself in all the Scriptures" (Luke 24:27).

It is prophecy that uniquely distinguishes the Bible from all other forms of "conventional" literature. Biblical prophecies are simple declarative statements about the future that have already been completely fulfilled, have been fulfilled in part, or are yet to be fulfilled. The intriguing challenge for those who study Bible prophecy is to know which is which. And that is where much of the controversy in Christian circles lies today—in squabbles about the proper interpretation of key prophetic passages, especially the relative timing and nature of Jesus' second coming, the Millennium, the Rapture, and the Great Tribulation. I have an opinion on these questions, but my purpose here is not to engage in those debates, at least not directly. Instead, I choose to focus on the fact that the Bible is demonstrably unique in that it contains declarative prophecy and that all such prophecy either has been or will be literally fulfilled as the future of this planet and the humanity that currently dwells upon it is played out.

In closing out this brief discussion, I believe it is worth highlighting the fact that the occurrence of demonstrably fulfilled prophecy within the pages of the Scripture is undeniable proof that the Bible is of divine origin and not the mere musings of imaginative men. I won't pause to list the hundreds of examples of all such fulfilled prophecies, but Lee Strobel's recent classic, *The Case for Christ,* is a good place for the interested reader to start.

THE BIBLE'S FIRST PROPHECY

The first prophecy of the Bible comes in the third chapter of Genesis, immediately after the famous failure of Adam and Eve to follow a simple rule, "from the tree of the knowledge of good and evil you shall not eat" (Gen. 2:17). While cursing the serpent who fooled Eve into first eating this forbidden fruit, God prophesied, "And I will put enmity between you and the woman, and between your seed and her seed; he shall bruise you on the head, and you shall bruise him on the heel" (Gen. 3:15). This active and mutual hatred is ultimately explained more fully in Revelation to have been a description of *the* spiritual struggle between Satan (the seed of the serpent) and Jesus (the seed of the woman), who is the virgin-born King of both an earthly nation, Israel, and an eternal heavenly kingdom, "the rest of her offspring, who keep the commandments of God and hold to the testimony of Jesus" (Rev. 12:17).

After giving this prophecy to the serpent, God confronted Eve and warned her with a prophecy about the personal struggles she and every woman after her have come to know all too well in this male-dominated world, "I will greatly multiply your pain in childbirth, in pain you shall bring forth children; yet your desire shall be for your husband, and he shall rule over you" (Gen. 3:16). At last, God prophesied to Adam, "Cursed is the ground because of you; in toil you shall eat of it all the days of your life. Both thorns and thistles it shall grow for you; and you shall eat the plants of the field; by the sweat of your face you shall eat bread, till you return to the ground, because from it you were taken; for you are dust, and to dust you shall return" (Gen. 3:17–19). Thus Adam was fired from his first job (tending the garden), and he set to work on the development of the world's oldest industry, agriculture. By the way, weed control (the "thorns and thistles") is still one of the greatest challenges to agriculture, and the environmental consequences of this have dominated much of my scientific career!

Here we must pause and ask a question about the true meaning and extent of "the curse." Over the years, some have taught that there was no biological death of any kind in any species on Earth prior to "the fall of man." In my years of studying the Bible, I have found no support or need for this particular interpretation. However, I do find a more compelling interpretation of the current state of the universe in the writings of Paul, "For we know that the whole creation groans and suffers the pains of childbirth together until now" (Rom. 8:22). In making this statement, Paul seems to merge the initial rebukes of Eve and Adam into one, and I think this is the proper interpretation of the original curse.

Regardless of what transpired before the appearance of *Homo sapiens* on Earth, our current plight is clear. We have overpopulated this globe to the extent that we unavoidably pollute the water we drink, the air we breathe, and the soil in which we sow our crops, "cursed is the ground because of us." Thus I interpret this to be an "environmental curse," in which our species is hurtling headlong into an inevitable collision between our exponentially-expanding combustion of fossil fuels and the limiting physical constraints of the nonlinear, coupled, natural systems of this planet. And it all began with one simple act of disobedience.

THE FLOOD

According to the Bible, ours is not the first time man has faced the curse of a coming global catastrophe from God. Ironically, it was a time of global warming, when the most recent period of heavy glaciation was coming to a natural end (most likely the sharp climb in global temperatures about 15,000 years ago, as shown in Figure 2.2). However, the global warming of

Noah's day was not man-made, at least not in the literal scientific sense of today's theory of man-made global warming. The world population was undoubtedly much smaller, and the combustion of fossil fuels could not have been as great as it is now, so far as we know. Nevertheless, man faced a near-term natural challenge analogous to ours.

God warned Noah, "And behold, I, even I am bringing the flood of water upon the earth, to destroy all flesh in which is the breath of life, from under heaven; everything that is on the earth shall perish" (Gen. 6:17). Subsequently, the Bible records the following account, "the fountains of the great deep burst open and the floodgates of the sky were opened. And the rain fell on the earth for forty days and forty nights" (Gen. 7:11–12). Now that's climate change! Especially for people who had not previously experienced *any* rain!

Of course, the story of a worldwide flood and Noah's successful execution of God's plan to rescue both his family and selected wildlife in an ark is one that modern skeptics are quick to wield as one of the first daggers against the reliability of the Bible. Today it is easy to step back and ask questions that could not have been asked directly some 3,500 years ago, when Moses is credited with having authored Genesis and the other four books of the Pentateuch: Exodus, Leviticus, Numbers, and Deuteronomy. The fact that there are very similar Flood stories in other world cultures, including people groups (such as Native Americans) very far from the Middle East, is sometimes used to argue that this story was simply borrowed and modified from extra-biblical sources, perhaps from a very similar ancient Babylonian legend. However, the same argument can be used in reverse, and I don't find this to be a credible critique.

I will not dwell on this particular debate, but instead I merely cite the climate record of the most recent millennia (shown earlier in Figure 2.2), which clearly shows a history of repeated glaciation phases and melting events, the most recent of which probably corresponds to the melting flood described in Genesis, beginning around 15,000 years ago (see Figure 2.2). Was this truly a worldwide flood that inundated every square inch of the Earth's land surface? The more skeptical scientist within me says no, especially given Schroeder's relativistic reconciliation of science with the six-day Creation account. But were there relatively catastrophic flooding episodes suffered by our distant ancestors in many world areas as the sea levels rose and certain inland seas were suddenly created when various glacial dams broke, especially around the Black Sea? I say yes, and I believe the available scientific data support such a conclusion.

More important to our current plight, however, is the eternal truth of the story of Noah. For those listening—"He who has ears, let him hear" (Matt. 13:9)—God is warning us now just as he warned Noah, "The end of all flesh has come before Me; for the earth is filled with violence because of them; and behold, I am about to destroy them with the earth" (Gen. 6:13). Jesus referred to a similar but drastically more catastrophic destruction when He prophesied of His own return:

> "Heaven and earth will pass away, but My words shall not pass away. But of that day and hour no one knows, not even the angels of heaven, nor the Son, but the Father alone. For the coming of the Son of Man will be just like the days of Noah. For as in those days which were before the flood they were eating and drinking, they were marrying and giving in marriage, until the day that Noah entered the ark, and they did not understand until the flood came and took them all away; so shall the coming of the Son of Man be" (Matt. 24:35–39).

Thus we see that all quibbling over the true nature of the Flood is mere prattle when confronting the reality of the Earth's coming fate.

Before proceeding further in the book of Genesis, it is also worth taking particular note of why the Earth was being destroyed in Noah's day—it was because of man, "the LORD saw that the wickedness of man was great on the earth, and that every intent of the thoughts of his heart was only evil continually" (Gen. 6:5). This stated rationale stands in stark contrast to Jesus' stated purpose for a second worldwide intervention, "And unless those days had been cut short, no life would have been saved; but for the sake of the elect those days shall be cut short" (Matt. 24:22). Unlike the days of Noah, when only he and "seven others" (2 Pet. 1:5) were saved, the Bible promises the ultimate deliverance of "a great multitude, which no one could count, from every nation and all tribes and peoples and tongues"

(Rev. 7:9). How many, exactly? Only God knows, but it would seem wise to enter the eternal ark of salvation (Jesus) while there is time.

CLIMATE AND THE BIRTH OF THE NATION ISRAEL

Following the Flood and throughout the remainder of the book of Genesis, the most significant climate event is that of repeated droughts and the famines that they brought. The first of these droughts was the cause for Abram's (later renamed Abraham) ill-fated journey out of Canaan to Egypt, "Now there was a famine in the land; so Abram went down to Egypt to sojourn there, for the famine was severe in the land" (Gen. 12:10). This first famine was just the first in a series of drought-induced famines that characterize the biblical record for Canaan, the land that subsequently became the earthly home for the nation of Israel. These droughts figured prominently in the family history of Abraham and his famous descendants: Isaac, Jacob (later renamed Israel), and his twelve sons, most notably Joseph.

The Genesis record also states that some of the droughts that struck the land of Canaan reached farther west, into the land of Egypt. When Joseph ended up in Egypt through the treachery of his brothers, he was able to correctly interpret a pair of Pharaoh's dreams as a warning of seven bountiful agricultural years to be followed by seven consecutive years of famine (presumably drought-induced). Pharaoh rewarded Joseph by placing him in a powerful position of authority throughout the land, subservient only to Pharaoh. As mentioned earlier, Joseph's story is a compelling foreshadowing of the eventual mission of Jesus, and his relationship to Pharaoh recalls the relationship of Jesus to the Father.

At any rate, when the predicted famine began, Joseph had wisely stored grain in secure facilities throughout the land of Egypt that brought Joseph's ten brothers back to kneel before him in a desperate search for food. The drought eventually drove the extended family of Joseph's father Jacob (just over seventy individuals) to leave the land of Canaan and dwell in the land of "Goshen." This Goshen is believed to have been the once fertile plain on the eastern portion of the Nile Delta. In just 430 years, the Seventy Israelites became "about six hundred thousand men on foot, aside from children" (Exod. 12:37). This rapid rate of growth might seem initially to be another "nonscientific" exaggeration of the kind that invites the ridicule of skeptics. However, it is actually a perfectly reasonable rate of population growth for such a fertile area and corresponds to a doubling time of about twenty-nine years.

As shown in Figure 3.1, this doubling time for the infant nation of Israel is roughly equal to the fastest observed doubling time for the *entire* world population, which was about thirty-two years in the early 1960s. Due to a number of factors, this doubling time has now slowed to about sixty years, but the number of people added annually to the world population is still growing. Currently, more than seventy-seven million people are being added every year. This annual increase is projected by the United Nations to peak very soon (in the year 2011) at about 78.5 million and then begin a slow decrease, with total world population eclipsing seven, eight, and nine billion in the years 2013, 2026, and 2042, respectively. It remains to be seen how accurate these projections are, but they certainly imply an increasing rate of global greenhouse gas emissions for the foreseeable future, particularly due to the rapid increase in per capita emissions in the two most populous nations of the world: China and India.

Plagues in Egypt

According to the first few verses of Exodus, the rapid growth of Jacob's (Israel's) descendants caused them to pose a potential political or military threat to their Egyptian hosts. The Hebrew Israelites then became the object of oppression from a new Pharaoh "who did not know Joseph" (Exod. 1:8). Pharaoh implemented a mandatory birth control program among the Israelite midwives that might even be offensive to modern abortionists, "When you are helping the Hebrew women to give birth and see them upon the birthstool, if it is a son, then you shall put him to death" (Exod. 1:16). It was during those first times of still-extant anti-Semitic racial persecution that Moses was hidden by his mother and lovingly placed in a wicker basket, only to be discovered by Pharaoh's daughter. Moses, thus saved, was raised as a prince of Egypt in the palace of Pharaoh but was later banished at age forty for killing an Egyptian whom Moses had seen beat a Hebrew slave. Forty years later, Moses was called back to Egypt by God to deliver the Israelites out of Egypt, thereby giving birth to the nation Israel—a nation that survives even today, despite an ever-present and escalating climate of racial hatred, whose roots lie deep in the jealous hearts of the nations that surround her. Moses and his older brother, Aaron, became witnesses for God before Pharaoh, pleading for the release of His people. But Pharaoh's heart was hardened as Moses and Aaron brought down increasingly severe plagues upon the land, foreshadowing the coming worldwide plagues detailed elsewhere throughout the books of the Prophets and especially in Revelation.

Although they don't refer specifically to climate, at least two of the ten Exodus plagues were weather related. In the seventh plague, hail and severe lightning struck the land of Egypt, with the sole exception that "the land of Goshen, where the sons of Israel were, there was no hail" (Exod. 9:26). In the ninth plague, Egypt was stricken by a thick "darkness" for three days, "even a darkness which may be felt" (Exod. 10:21). Some of the other plagues that befell Egypt are themselves usually triggered by particular environmental or weather conditions of heat, drought, or excessive humidity (red tide, frogs, gnats, locusts, and diseases among both livestock and humans), but no natural or other weather-related causes are mentioned explicitly in the text.

As most will recall, the last of the ten plagues is described as being entirely supernatural, in which God personally descended to perform a seemingly horrific act, "the LORD struck all of the first-born in the land of Egypt, from the first-born of Pharaoh who sat on his throne to the first-born of the captive who was in the dungeon, and all the first-born of the cattle" (Exod. 12:29). The Exodus record also states that God "passed over" the homes of the Israelites, who had each sacrificed a lamb without blemish and painted a bit of the blood on the post and lintel of the central doorway to their homes.

This Passover has been celebrated ever since by the Jewish people, and it reached its first historical fulfillment in the sacrifice made by Jesus, who was described by John the Baptist as "the Lamb of God, who takes away the sins of the world!" (John 1:29). His own blood painted the "post and lintel" of the cross on the first day of a Passover nearly 2,000 years ago. But it also points forward to a future Passover, in which those chosen by God will receive a supernatural seal or mark (Rev. 7:3 and Rev. 9:4) that will enable them to miraculously avoid a future worldwide pandemic in which victims will be tormented "for five months; and their torment was like the torment of a scorpion when it stings a man. And in those days men will seek death and will not find it; and they will long to die and death flees from them" (Rev. 9:5–6).

But this brief digression on the last and most horrendous of the ten Egyptian plagues of Moses' day has taken us away from the study of climate in the Bible, and we must return. There were forty years of wandering in the desert and then 400 years of battles as the descendants of Israel fought their way back into the land of Canaan. These four centuries are known as the time of the Judges, named after the historical book that describes this period, one marked by much war and little mention of climate or weather. A pleasant exception is the story of Ruth.

Drought in the Story of Ruth

I find the book of Ruth to be the Bible's most touching and compelling, nestled between the somewhat dark book of Judges and the more famous tales of King Saul and King David that are told in first and second Samuel. The circumstance that drove the entire plot was that of a drought-induced famine, "Now it came about in the days when the judges governed, that there was a famine in the land. And a certain man of Bethlehem in Judah went to sojourn in the land of Moab with his wife and two sons" (Ruth 1:1). The man's name was Elimelech and his wife's name was Naomi, and their two Jewish sons took Moabite wives. As invariably happened when any of the Israelites ventured away from the Land of Promise, Naomi's life was marred by tragedy in the foreign land when her husband and both her sons died in the land of Moab.

But another change of local climate caused Naomi to move again, so that, "she arose with her daughters-in-law that she might return from the land of Moab, for she had heard in the land of Moab that the LORD had visited His people in giving them food" (Ruth 1:6). One of many emotional moments in the story occurred when Ruth, one of Naomi's two daughters-in-law, refused to heed Naomi's plea that the daughters-in-law remain in Moab, rather than making the long trek back to Israel. Ruth famously pledged to bind her life to that of her mother-in-law, "where you go, I will go; and where you lodge, I will lodge. Your people shall be my people, and your God my God. Where you die, I will die, and there I will be buried. Thus may the Lord do to me, and worse, if anything but death parts you and me" (Ruth 1:16–17). The book then becomes a simple and tender love story as Ruth is then taken as a wife by one of her husband's near kinsmen, Boaz. Their first child was a son, Obed, who became the grandfather of Israel's most beloved King, David—an apt name, as it is Hebrew for the word, "beloved."

Israel's Psalmists Sing of God's Sovereignty over Climate

King David, despite his notorious and very human failings, was uniquely described by God, in Paul's first recorded sermon (at Pisidian Antioch), as being "a man after Mine own heart, which shall fulfill all My will" (Acts 13:22, KJV). David's many military exploits are chronicled in great detail within the historical books of the Bible, but his direct literary contribution is contained in the book of Psalms with about half of its 150 chapters either directly attributed to or in some way associated with him. The Psalms sink to the depths and soar to the heights of human emotion, a testament to the fact that David had no problem putting the deepest longings of his heart into words. Along the way, David and the other contributors to the Psalms poetically painted the picture of a God who is intimately involved with the Earth's climate. Some of the numerous climate-related passages in the Psalms are listed below:

- *"From the brightness before Him passed His thick clouds, hailstones and coals of fire. The LORD also thundered in the heavens, and the Most High uttered His voice, hailstones and coals of fire"* (Ps. 18:12–13).
- *"He gathers the waters of the sea together as a heap; He lays up the deeps in storehouses"* (Ps. 33:7).
- *"Thou dost water its furrows abundantly; Thou dost settle its ridges; Thou dost soften it with showers; Thou dost bless its growth"* (Ps. 65:10).
- *"Thou didst shed abroad a plentiful rain, O God; Thou didst confirm Thine inheritance, when it was parched"* (Ps. 67:9).
- *"Thou didst break open springs and torrents; Thou didst dry up ever-flowing streams"* (Ps. 74:15).

- "The sound of Thy thunder was in the whirlwind; the lightnings lit up the world; the earth trembled and shook" (Ps. 77:18).
- "He caused the east wind to blow in the heavens; and by His power He directed the south wind" (Ps. 78:26).
- "Thou dost rule the swelling of the sea; when its waves rise, Thou dost still them" (Ps. 89:9).
- "He waters the mountains from His upper chambers; the earth is satisfied with the fruit of His works. He causes the grass to grow for the cattle, and vegetation for the labor of man, so that he may bring forth food from the earth, and wine which makes man's heart glad, so that he may make is face glisten with oil, and food which sustains man's heart" (Ps. 104:13–15).
- "Those who go down to the sea in ships, who do business on great waters; they have seen the works of the Lord, and His wonders in the deep. For He spoke and raised up a stormy wind, which lifted up the waves of the sea" (Ps. 107:23–25).
- "He caused the storm to be still, so that the waves of the sea were hushed" (Ps. 107:29).
- "He changes rivers into a wilderness [desert], and springs of water into a thirsty ground; a fruitful land into a salt waste, because of the wickedness of those who dwell in it. He changes a wilderness [desert] into a pool of water, and a dry land into springs of water; and there He makes the hungry to dwell, so that they may establish an inhabited city, and sow fields, and plant vineyards, and gather a fruitful harvest" (Ps. 107:33).
- "Whatever the Lord pleases, He does, in heaven and in earth, in the seas and in all deeps. He causes the vapors to ascend from the ends of the earth; Who makes lightnings for the rain; Who brings forth the wind from His treasuries" (Ps. 135:6–7).
- "Sing to the Lord with thanksgiving; sing praises to our God on the lyre, who covers the heavens with clouds, who provides rain for the earth, who makes grass to grow on the mountains" (Ps. 147:7–8).
- "He gives snow like wool; He scatters the frost like ashes. He cast forth His ice as fragments; who can stand before His cold? He sends forth His word and melts them; He causes His wind to blow and the waters to flow" (Ps. 147:16–18).

Taken in their entirety, the Psalms expressed what had become a widely held view among the Jewish people that God utilized weather and climate to express His view about man's behavior. This linkage was made even more explicit during the life and times of David's immediate successor, Solomon.

SOLOMON'S TEMPLE DEDICATION PRAYER AND NIGHT VISION

David had numerous children through his many wives and concubines, and therefore many possible heirs to his throne. Through a bit of royal court intrigue explained in 1 Kings 1, the scepter passed to Solomon, whose mother (Bathsheba) David had married through the darkest act of his life (2 Samuel 11), a treacherous crime involving both adultery and cold-blooded murder, all orchestrated via an unseemly exercise of his power as King.

Whether due to guilt over these sins, or from a true heart of worship (only God knows), David had it in his heart to build a massive temple to honor God. However, the task fell to Solomon, shortly after his ascendancy to the throne. At the dedication ceremony for the First Temple, Solomon proclaimed a public prayer—obviously, the ACLU hadn't yet gained their current power to stop such dangerous acts—in which the sovereignty of God over weather was explicitly acknowledged, "When the heavens are shut up and there is no rain because they have sinned against Thee, and they pray toward this place and confess Thy name, and turn from their sin when Thou dost afflict them; then hear Thou in heaven and forgive the sin of Thy servants and Thy people Israel, indeed, teach them the good way in which they should walk. And send rain on Thy land, which Thou hast given to Thy people for an inheritance" (2 Chron. 6:26–27).

That same night, God visited Solomon in a dream and acknowledged what Solomon had prayed and made him a promise, "If I shut up the heavens so that there is no rain, or if I command the locust to devour the land, or if I send pestilence among My people, and My people who are called by My name humble themselves and pray, and seek My face and turn from their wicked ways, then I will hear from heaven, will forgive their sin, and will heal their land" (2 Chron. 7:13–14). This passage helped confirm the very close connection that the nation Israel felt existed between God's assessment of the nation's spiritual behavior and the weather patterns affecting the country. This may seem unscientific, but in reality there is no way to evaluate the question of causality from a scientific perspective—because the actual weather pattern at any point on the planet is completely unpredictable and beyond the realm of science to diagnose. This was explained in Chapter 2, during the discussion of Lorenz's Butterfly Effect. It may seem trite, but it's actually true—only God knows whether it will actually rain where you are now two weeks from Tuesday—and only He knows why.

This Jewish attitude associating Deity with the ability to control weather carried right on through to the days when Jesus was born into the nation of Israel. So this helps explain why Jesus' disciples took an incident of His control of weather as a direct demonstration of His deity. They had entered into a boat with Him to row across the Sea of Galilee, "And there arose a fierce gale of wind, and the waves were breaking over the boat so much that the boat was already filling up. And He Himself was in the stern, asleep on the cushion; and they awoke Him and said to Him, 'Teacher, do You not care that we are perishing?' And being aroused, He rebuked the wind and said to the sea, 'Hush, be still.' And the wind died down and it became perfectly calm. And He said to them, 'Why are you so timid? How is it that you have no faith?' And they became very much afraid and said to one another, 'Who then is this, that even the wind and the sea obey Him?'" (Mark 4:37–41).

Elijah: In with a Drought and "Up, Up, and Away" in a Whirlwind

But there was another prophet about 700 years before Jesus, Elijah, who also dabbled in "climate control." In fact, the very first Biblical mention of Elijah contains a remarkably brash statement from the prophet, "Now Elijah the Tishbite, who was of the settlers of Gilead, said to Ahab, 'As the LORD, the God of Israel lives, before whom I stand, surely there shall be neither dew nor rain these years, except by my word.'" (1 Kings 17:1).

As you can see, Elijah claimed not only an ability to stop the rain, but he also boasted of an ability to prevent the dew as well. Later, after his thorough drubbing of the prophets of Baal in the battle of Mt. Carmel, Elijah decided it was time for the rains to return, "Now Elijah said to Ahab, 'Go up, eat and drink; for there is the sound of the roar of a heavy shower.'"

(1 Kings 18:41). And he was right about that prediction as well. Not bad for a mere mortal. In fact, James (Jesus' half brother) reminds us of Elijah's entirely human nature when he later wrote of him, "Elijah was a man with a nature like ours, and he prayed earnestly that it might not rain: and it did not rain on the earth for three years and six months. And he prayed again, and the sky poured rain, and the earth produced its fruit" (James 5:17–18).

It's of interest to note that this unique (within the Old Testament) and divinely bequeathed power of Elijah to impact rainfall is mirrored in the capabilities of the mysterious Two Witnesses of Revelation, who "have the power to shut up the sky, in order that rain may not fall during the days of their prophesying" (Rev. 11:6). The two witnesses have other powers that seem reminiscent of Moses and Aaron at the time of the Exodus, "power over waters to turn them into blood, and to smite the earth with every plague, as often as they desire" (Rev. 11:6). This has led some commentators to equate these Two Witnesses with Elijah and Moses, which seems particularly compelling considering their co-appearance with Jesus at the Mount of Transfiguration, but one nasty detail to reconcile is the fact that many older manuscripts repeatedly declare that the two witnesses share only one body. Logic would suggest that the manuscripts with "body" instead of "bodies" are more likely correct, as it would seem tempting for a copyist to correct the apparent discrepancy in number. Hmmmm...

But getting back to Elijah, himself, the linkages to weather phenomena in his life and times go beyond drought. One reasonable interpretation of Elijah's victory at Mt. Carmel is that he was somehow able to have God to send down lightning upon his verbal command. Not too long after this quite literal "mountaintop experience," Elijah was found hiding for his life in a cave on yet another mountaintop, this time Mt. Horeb (where Moses had first met the Lord). There God scolded Elijah for his sudden lack of courage and taught Elijah a little lesson about Himself:

> "So He said, 'Go forth, and stand on the mountain before the LORD.' And behold, the LORD was passing by! And a great and strong wind was rending the mountains and breaking in pieces the rocks before the LORD; but the LORD was not in the wind. And after the wind was an earthquake, but the LORD was not in the earthquake. And after

the earthquake a fire, but the LORD was not in the fire; and after the fire a sound of gentle blowing. And it came about when Elijah heard it, that he wrapped his face in his mantle, and went out and stood in the entrance of the cave. And behold, a voice came to him and said, 'What are you doing here, Elijah?'" (1 Kings 19:11–13).

Thus God taught Elijah to listen for His "still, small voice" in the gentlest of breezes, which seems to echo the refrain of Bob Dylan's familiar song, "the answer, my friend, is blowing in the wind." At any rate, God's specific message to Elijah involved a journey to Damascus during which Elijah was to nominate his successor, Elisha, who went on to become the closest eyewitness to the final earthly weather event that accompanied the life of Elijah when the prophet "went up by a whirlwind to heaven" (2 Kings 2:11).

GOD ANSWERS JOB "OUT OF THE WHIRLWIND"

The ancient book of Job (perhaps the Bible's oldest), marks the end of the historical narrative books. Job contains a number of references to weather, including a weather observation that still seems to hold true for the Northern Hemisphere, "Out of the south cometh the whirlwind" (Job 37:9 KJV). A severe weather event played a central role in establishing the basic plot of Job, when "a great wind came from across the wilderness and struck the four corners of the house, and it fell on the young people and they died" (Job 1:19). The "young people" of this passage were Job's seven sons and three daughters, and apparently all of their young families. The news of this tragedy was the last of four rapid-fire catastrophic reports to reach the ears of Job, which caused him to fall to his knees, worship God, and mournfully pray, "Naked I came from my mother's womb, and naked I shall return there. The LORD gave, and the LORD has taken away; blessed be the name of the LORD" (Job 1:21).

God's power over the weather and man's inability to comprehend it are highlighted later in the book during the discourse of Elihu, "He draws up the drops of water, they distill rain from the mist, which clouds pour down, they drip upon Man abundantly. Can anyone understand the spreading of the clouds, the thundering of His pavilion?" (Job 36:27–29). As we now know scientifically from the work of Lorenz on the "Butterfly Effect," the answer to Elihu's question is "No!" Later, Elihu continued, "God thunders with His voice wondrously, doing great things which we cannot comprehend. For to the snow He says, 'Fall on the earth,' and to the downpour and the rain, 'Be strong'" (Job 37:5–6). Elihu went on to say this, "From the breath of God ice is made, and the expanse of the waters is frozen. Also with moisture He loads the thick cloud; He disperses the cloud of His lightning. And it changes direction, turning around by His guidance, that it may do whatever He commands it on the face of the inhabited earth" (Job 37:10–12). This entire passage reaffirms the deeply held Jewish tradition that the sovereignty of God extends to the weather.

The final mention of weather-related terminology in Job comes when God's patience with the relentless ranting of Job's friends has finally run its course when, "the LORD answered Job out of the whirlwind" (Job 38:1). God's "answer" is actually a series of sixty-four unanswerable questions that are meant to convey the seemingly unfathomable gap between man and God and pointing out the ultimate folly of ever having the prideful notion of asking God the "Why?" question.

JESUS' COMMENTS ON THE WEATHER

Shifting to the New Testament, in the Sermon on the Mount, Jesus says, "You have heard that it was said, 'You shall love your neighbor, and hate your enemy.' But I say to you, love your enemies, and pray for those who persecute you in order that you may be sons of your Father who is in heaven; for He causes His sun to rise on the evil and the good, and sends rain on the righteous and the unrighteous" (Matt. 5:43–45).

Jesus' comment about man's ability to predict weather is a very interesting one, "When it is evening, you say, 'It will be fair weather, for the sky is red.' And in the morning, 'There will be a storm today, for the sky is red and threatening.' Do you know how to discern the appearance of the sky, but cannot discern the signs of the times?" (Matt. 16:2–3). This seems a particularly relevant question to ask of modern man.

GOD'S USE OF "STURM" UND "DRANG"

This colorful Germanic term for "storm and stress" helps conjure up a powerful visual image of the several intense storms in the Bible. We have already reviewed the incident on the Sea of Galilee, when Jesus sent great fear into the hearts of his disciples by displaying His ability to calm the wind and waves through the seemingly effortless utterance of the two words, "be still." At least two other sea storms figure prominently in the Bible, both on the Mediterranean, which the Jewish people seemed not meant to sail. The first of these came in the book of Jonah, in which the reluctant prophet foolishly tried to flee God's call to the wicked city of Nineveh. Jonah tried to escape by boat, but God sent a great storm, which the superstitious shipmates of Jonah discover (by drawing lots) must be due to Jonah's disobedience. Proving this, they threw Jonah overboard and the storm stopped. But Jonah's trouble had only started, when a fish then swallowed him whole. Three days in the fish's acidic belly finally broke Jonah's stubborn will, so that he finally obeyed God and carried the message of repentance, "Yet forty days, and Nineveh will be overthrown" (Jonah 3:4). Parenthetically, these forty days are eerily reminiscent of the forty unrepentant years after the crucifixion of Jesus that the mainstream Jews of Jerusalem lived until that city was overthrown. The parallel is made all the stronger by the fact that Jesus pointed to the story of Jonah whenever the Pharisees asked him for a "sign" or miracle to convince them of the truth of His message.

Strangely, and with a profound message for those of us called to bear God's warning to this modern world, Jonah sulked like a pouting four-year old after Nineveh heeded the warning and did repent. After God sent a withering east wind and a blazing sun that caused Jonah to nearly faint, God asked the pouting prophet, "should I not have compassion on Nineveh, the great city in which there are more than 120,000 persons who do not know the difference between their right and left hand, as well as many animals?" (Jonah 4:11). My humble prayer is that I would never have the same lack of mercy in my own heart while being called to warn this world of the peril that appears to lie in wait for us.

The other great sea storm is one that marks the last historical incident recorded in the Bible, Paul's ill-fated ship journey from a Caesarea prison to Rome, where he would await his final trial and certain execution. Along the way, the boat was caught in "a violent wind, called Euraquilo" (Acts

27:14). The ship was caught by the wind and driven through a frightful storm for many days, which gave Paul ample opportunity to display God's power over the elements when he bravely swore to those who thought they were in charge of the ship:

> "Men, you ought to have followed my advice and not to have set sail from Crete, and incurred this damage and loss. And yet now I urge you to keep up your courage, for there shall be no loss of life among you, but only of the ship. For this very night an angel of the God to whom I belong and whom I serve stood before me, saying, 'Do not be afraid, Paul; you must stand before Caesar; and behold, God has granted you all those who are sailing with you.' Therefore, keep up your courage, men, for I believe God, that it will turn out exactly as I have been told. But we must run aground on a certain island" (Acts 27:21–25).

Sure enough, the ship was run aground two weeks later, upon the shores of the island of Malta, and every member of the ship survived the entire journey just as Paul had sworn, all 276 of them. I am tempted to draw a parallel here between this final historical narrative of the Bible and the coming storms that will soon engulf this Good Ship Planet Earth, if the modern climate scientists are right. I believe there are rough seas ahead, but every human soul that has ever lived is known by the Creator by name, and He shall pluck each from the sea of time. Some will answer for their wanton lives of callous unbelief, and the others will enter into the eternal joy of serving their Lord and Master forever. I thank Him and praise Him forever that He chose to include me in the latter group—it was totally without any merit on my part.

Now that we have completed an overview of the climate events of the past, as recorded in Scripture, it is time to consider what the Bible has to say about climate events of the future. For this, we must dive headlong into prophecy and the challenging concept of predestination—in which Scripture teaches that not only are certain individuals called to be prophets before they were even born, but the ultimate fate of the entire planet has been predetermined by God—before He created it.

The Call of the Prophet, Predestiny Beckons

As we just reviewed in the story of Jonah, God calls His servants, the prophets, and His will is ultimately done, despite the reluctance the called prophet nearly always exhibits. For instance, when Moses was called he tried to use his speech impediment as an excuse, "I am slow of speech and slow of tongue" (Exod. 4:10). Somewhat humorously, God reminded Moses that his brother would be quite a good spokesman for the people, "Is there not your brother Aaron the Levite? I know that he speaks fluently" (Exod. 4:14). Thus was formed the first family ministry, soon to be joined by their sister, Miriam.

The deeply mysterious truth of the Bible is that the very words it contains were recorded by human beings chosen by God, whom we know as prophets. When were they chosen? According to Scripture, long before they were born. David wrote of this notion of predestination, "My frame was not hidden from Thee when I was made in secret, and skillfully wrought in the depths of the earth. Thine eyes have seen my unformed substance; and in Thy book they were all written, the days that were ordained for me, when as yet there was not one of them" (Ps. 139:15–16). The same idea marks the opening passages of Jeremiah, "Before I formed you in the womb I knew you, and before you were born I consecrated you; I have appointed you a prophet to the nations" (Jer. 1:5). For the prophet, disobedience was not an option, certainly not a pleasant one, as explained by Ezekiel, "Now as for you, son of man, I have appointed you a watchman for the house of Israel; so will hear a message from My mouth, and give them warning from Me. When I say to the wicked, 'O wicked man, you shall surely die,' and you do not speak to warn the wicked from his way, that wicked man shall die in his iniquity, but his blood I will require from your hand" (Ezek. 33:7–8).

Paul, perhaps the most intellectually gifted of all the biblical authors, took the doctrine of predestination even further, "For whom He foreknew, He also predestined to become conformed to the image of His Son, that He might be the firstborn among many brethren; and whom He predestined, these He also called: and whom He called, these He also justified: and whom He justified, these He also glorified" (Rom. 8:29–30).

When Israel's prophets wrote of God and weather, they affirmed God's active control and the close connection between weather and the nation's spiritual health, as reflected in the passages listed below:

- *"When He utters His voice, there is a tumult of waters in the heavens, and He causes the clouds to ascend from the end of the earth; he makes lightning for the rain, and brings forth the wind from His storehouses"* (Jer. 51:16).
- *"I withheld the rain from you while there were still three months until harvest. Then I would send rain on one city and on another city I would not send rain"* (Amos 4:7).
- *"God appointed a scorching east wind, and the sun beat down on Jonah's head so that he became faint and begged with all his soul to die"* (Jonah 4:8).
- *"The LORD is slow to anger and great in power, and the LORD will be no means leave the guilty unpunished. In whirlwind and storm is His way, and clouds are the dust beneath His feet. He rebukes the sea and makes it dry; He dries up all the rivers"* (Nah. 1:3–4).
- *"I smote you and every work of your hands with blasting wind, mildew, and hail"* (Hag. 2:17).
- *"Ask rain from the LORD at the time of the spring rain – the LORD who makes the storm clouds; and He will give them showers of rain, vegetation in the field to each man"* (Zech. 10:1).

False Prophets Chosen Too!

But there is a dark side to the idea of predestination that is very difficult for modern man to swallow—namely this—that God, in not choosing some to respond, has implicitly chosen others to endure the wrath of death that is the payment due for their lives of sin. Jude spoke of this dark doctrine when he wrote, "For certain persons have crept in unnoticed, those who were long beforehand marked out for this condemnation, ungodly persons who turn the grace of our God into licentiousness and deny our only Master and Lord, Jesus Christ" (Jude 4).

Jeremiah was a prophet in a time such as ours as well, whose testimony completely contradicted the false prophets of his day, about whom he wrote,

"Thus says the LORD of hosts, 'Do not listen to the words of the prophets who are prophesying to you. They are leading you into futility; they speak a vision of their own imagination, not from the mouth of the LORD. They keep saying to those who despise Me, "The LORD has said, 'You will have peace'"; and as for everyone who walks in the stubbornness of his own heart, they say, "Calamity will not come upon you." But who has stood in the council of the LORD, that he should see and hear His word? Who has given heed to His word and listened? Behold the storm of the LORD has gone forth in wrath, even a whirling tempest; it will swirl down on the head of the wicked. The anger of the LORD will not turn back until He has performed and carried out the purposes of His heart; in the last days you will clearly understand it" (Jer. 23:16–20).

Peter also wrote of a mocking attitude, "Know this first of all, that in the last days mockers will come with their mocking, following after their own lusts, and saying, 'Where is the promise of His coming? For ever since the fathers fell asleep, all continues just as it was from the beginning of creation'" (2 Pet. 3:3–4). Paul warned of the widespread popularity of the false prophets that are now among us, "For the time will come when they will not endure sound doctrine; but wanting to have their ears tickled, they will accumulate for themselves teachers in accordance to their own desires; and will turn away their ears from the truth, and will turn aside to myths" (2 Tim. 4:3–4).

So who are the false prophets of our day? It is my contention that these are the ones who deny the reality of the prophetic passages of the Bible—that this Earth will experience increasingly severe climatic disruptions until it is ultimately destroyed with only one way of escape available to all of humanity, "I am the way, and the truth, and the life; no ones comes to the Father but through Me" (John 14:6). As for the majority of the world population, which clearly recoils against this "intolerant" teaching of Jesus, the writings of Solomon, as given in the book of Proverbs, speak clearly and with a stinging relevance despite the three millennia that have passed since the words were first recorded:

- *"The fear of the LORD is the beginning of knowledge; fools despise wisdom and instruction"* (Prov. 1:7).
- *"Wisdom shouts in the street, she lifts her voice in the square"* (Prov. 1:20).
- *"How long, O naive ones, will you love simplicity? And scoffers delight themselves in scoffing, and fools hate knowledge?"* (Prov. 1:22).
- *"Because I called and you refused; I stretched out my hand and no one paid attention; and you neglected all my counsel, and did not want my reproof; I will even laugh at your calamity. I will mock when your dread comes, when your dread comes like a storm, and your calamity comes on like a whirlwind, when distress and anguish come on you"* (Prov. 1:24–27).

No Excuses!

Man is a clever creature, and quick to shift blame to someone else for his own actions. Adam displayed this trait very adroitly when God first confronted him about eating the forbidden fruit, "The woman whom Thou gavest to be with me, she gave me from the tree, and I ate" (Gen. 3:12). Thus in one quick sentence he tried to blame both Eve and God for his own act of disobedience! Ever since that first denial, man has taken the art of ducking personal responsibilities to ever more sophisticated levels. We've created entire branches of the social sciences in a vain attempt to do away with the concept of sin.

But the Bible is clear that man is entirely without any legitimate excuse when it comes to the question of whether God really exists and whether we are really as sinful as Scripture says we are. In the opening chapter of Romans, quite possibly the most magnificent of all the books of the Bible, Paul doesn't pull any punches as he pronounces the list of accusations against all of humanity, the most grievous of which is that we "suppress the truth in unrighteousness" (Rom. 1:18). Paul also argues that man cannot claim ignorance as a defense, because "since the creation of the world His invisible attributes, His eternal power and divine nature, have been clearly seen, being understood through what has been made, so that they are without excuse" (Rom. 1:20).

In our day, all of Paul's arguments are that much stronger, as we continue to collect a vast amount of scientific data pointing to the irreducibly complex nature of the biological systems and the exquisite balance of physical forces in the universe that have been thoughtfully and powerfully crafted to make the life we enjoy on this planet possible. Indeed, Daniel prophesied of our days of globalization and exploding increases in knowledge, "But as for you, Daniel, conceal these words and seal up the book until the end of time; many will go back and forth and knowledge will increase" (Dan. 12:4).

In addition to the vast increases in the availability of secular information, we also live in a day when the Bible itself is more accessible than ever before and in many more forms (radio, TV, podcasts, etc.) than ever before, as prophesied by Isaiah, "the earth will be full of the knowledge of the LORD as the waters cover the sea" (Isa. 11:9). Thus we have no one else to blame if we find ourselves duped into believing the many false prophets among us.

What is Man's Role in the Fate of the Earth?

From the very outset, God determined that mankind would have dominion over the Earth. This is made clear in the first chapter of the Bible, "Let Us make man in Our image, according to Our likeness; and let them rule over the fish of the sea and over the birds of the sky and over the cattle and over all the earth, and over every creeping thing that creeps on the earth. Be fruitful and multiply, and fill the earth, and subdue it; and rule over the fish of the sea, and over the birds of the sky, and over every living thing that moves on the earth" (Gen. 1:26–28). This gift of the Earth to all of mankind is praised in the Psalms, "What is man, that Thou dost take thought of him? And the son of man, that Thou does care for him? Yet Thou hast made him a little lower than God, and dost crown him with glory and majesty! Thou dost make him to rule over the works of Thy hands; Thou hast put all things under his feet, all sheep and oxen, and also the beasts of the field, the birds of the heavens, and fish of the sea, whatever passes through the paths of the seas" (Ps. 8:4–8). The same sentiment is reaffirmed later in the same book, "The heavens are the heavens of the LORD; but the earth He has given to the sons of men" (Ps. 115:16).

But what have we done with this gift? I fear that we have not treated the Earth kindly. If the theory of man-made global warming is true, it is clear that our treatment has rapidly gone from being merely unkind to being downright brutal. Indeed, the Bible teaches that man's actions are bringing about the literal destruction of the Earth, "the time has come...to destroy those who destroy the earth" (Rev. 11:18). This echoes the curse initially proclaimed upon the Earth due to Adam's disobedience, "Cursed is the ground because of you" (Gen. 3:17).

This basic thesis of mankind's direct culpability in the Earth's plight is entirely consistent with Moses' final admonitions to the sons of Israel as they were about to enter the promised land, when he warned them of the series of very negative consequences if they were disobedient to the Law, "The LORD will smite you with consumption and with fever and with inflammation and with fiery heat and with drought and with blight and with mildew, and they shall pursue you until you perish. And the heaven which is over your head shall be bronze, and the earth which is under you, iron. The Lord will make the rain of your land powder and dust; from heaven it shall come down on you until you are destroyed" (Deut. 28:22–24). These are only three verses within a series of over fifty verses of curses that God rains down upon the people from the mouth of Moses in his last days upon the Earth.

There are numerous similar passages from all of the Prophets. One penned by Isaiah is particularly relevant to the present discussion, "The earth is also polluted by its inhabitants, for they transgressed laws, violated statutes, broke the everlasting covenant. Therefore a curse devours the earth, and those who live in it are held guilty. Therefore the inhabitants of the earth are burned, and few men are left" (Isa. 24:5–6). And what is the historical record of the behavior of Israel? They were disobedient, just as all of man has been from Adam. Thus the destruction that is now coming to this planet and the humanity upon it is entirely consistent with the consequences that God had promised.

THE CERTAINTY OF EARTH'S DESTRUCTION

Peter speaks of this near the end of his second letter:

> "the heavens will pass away with a roar and the elements will be destroyed with intense heat, and the earth and its works will be burned up. Since all these things are to be destroyed in this way, what sort of people ought you to be in holy conduct and godliness,

looking for and hastening the coming of the day of God, on account of which heavens will be destroyed by burning, and the elements will melt with intense heat! But according to His promise we are looking for new heavens and a new earth, in which righteousness dwells" (2 Pet. 3:10–13).

This is consistent with how John the Baptist described the coming mission of Jesus, "As for me, I baptize you water; but One is coming who is mightier than I, and I am not fit to untie the thong of his sandals; He will baptize you with the Holy Spirit and fire. And his winnowing fork is in His hand to thoroughly clear His threshing floor, and to gather the wheat into His barn; but he will burn up the chaff with unquenchable fire" (Luke 3:16–17).

Jesus later shed further light on this imagery when he gave the parable of the tares and the wheat:

> "The kingdom of heaven may be compared to a man who sowed good seed in his field. But while men were sleeping, his enemy came and sowed tares also among the wheat and went away. But when the wheat sprang up and bore grain, then the tares became evident also. And the slaves of the landowner came and said to him, 'Sir, did you not sow good seed in your field? How then does it have tares?' And he said to them, 'An enemy has done this!' And the slaves said to him, 'Do you want us, then, to go and gather them up?' But he said, 'No; lest while you are gathering up the tares; you may root up the wheat with them. Allow both to grow together until the time of the harvest; and in the time of the harvest I will say to the reapers, "First gather up the tares and bind them in bundles to burn them up; but gather the wheat into my barn"'" (Matt. 13:24–30).

Jesus subsequently went on to explain this particular parable to the disciples with very explicit clarity,

> "The one who sows is the Son of Man, and the field is the world; and as for the good seed, these are the sons of the kingdom; and

the tares are the sons of the evil one; and the enemy who sowed them is the devil, and the harvest is the end of the age; and the reapers are angels. Therefore just as the tares are gathered up and burned with fire, so shall it be at the end of the age. The Son of Man shall send forth His angels, and they will gather out of His Kingdom all stumbling blocks, and those who commit lawlessness, and will cast them into the furnace of fire; in that place there shall be weeping and gnashing of teeth. Then the righteous will shine forth as the sun in the kingdom of their Father. He who has ears, let him hear" (Matt. 13:37–43).

Sometimes critics claim that these frequently cited end-times passages from Peter's letter and from Matthew are isolated in nature and inconsistent with what the rest of the Bible has to say about the fate of Earth. But similar language about the certainty of Earth's destruction is peppered throughout the Bible, and it is clear that the Kingdom of which Jesus spoke is not of this world and that we are His ambassadors from another "country":

- *"Ask of Me, and I will surely give the nations as Thine inheritance, and the very ends of the earth as Thy possession. Thou shalt break them with a rod of iron, Thou shalt shatter them like earthenware"* (Ps. 2:8–9).
- *"You will make them as a fiery oven in the time of Your anger; the LORD will swallow them up in His wrath, and fire will destroy them. Their offspring Thou wilt destroy from the earth, and their descendants from among the sons of men"* (Ps. 21:9).
- *"He will sweep them away with a whirlwind, the green and the burning alike"* (Ps. 58:9).
- *"For a cup is in the hand of the LORD, and the wine foams; it is well mixed, and He pours out of this; surely all the wicked of the earth must drain and drink down its dregs"* (Ps. 75:8).
- *"Of old Thou didst found the earth; and the heavens are the work of Thy hands. Even they will perish, but Thou dost endure; and all of them will wear out like a garment; like clothing Thou wilt change them, and they will be changed. But Thou art the same, and Thy years will not come to an*

end. The children of Thy servants will continue, and their descendants will be established before Thee" (Ps. 102:25–28).
- "For behold, I create new heavens and a new earth; and the former things shall not be remembered or come to mind" (Isa. 65:17).
- "'For just as the new heavens and the new earth which I make will endure before Me,' declares the LORD, 'So your offspring and your name will endure. And it shall be from new moon to new moon and from sabbath to sabbath, all mankind will come to bow down before Me,' says the LORD. 'Then they shall go forth and look on the corpses of the men who have transgressed against Me. For their worm shall not die, and their fire shall not be quenched; and they shall be an abhorrence to all mankind'" (Isa. 66:22–24).
- "'Behold, I am against you, O destroying mountain, who destroy the whole earth,' declares the LORD, 'And I will stretch out My hand against you, and roll you down from the crags and I will make you a burnt out mountain'" (Jer. 51:25).
- "Then the LORD, my God, will come, and all the holy ones with Him! And it will come about in that day that there will be no light; the luminaries will dwindle. For it will be a unique day known to the LORD, neither day nor night, but it will come about that at evening time there will be light" (Zech. 14:7).
- "For behold, the day is coming, burning like a furnace; and all the arrogant and every evildoer will be chaff; and the day that is coming will set them ablaze" (Mal. 4:1).
- "In My Father's house are many dwelling places; if it were not so, I would have told you; for I go to prepare a place for you" (John 14:2).
- "My kingdom is not of this world. If My kingdom were of this world, then My servants would be fighting, that I might not be delivered up to the Jews; but as it is, My kingdom is not of this realm" (John 18:36).
- "Therefore, we are ambassadors for Christ, as though God were entreating through us; we beg you on behalf of Christ, be reconciled to God" (2 Cor. 5:20).
- "By faith Abraham, when he was called, obeyed by going out to a place which he was to receive for an inheritance; and he went out, not knowing where he was going. By faith he lived as an alien in the land of promise, as in a foreign land, dwelling in tents with Isaac and Jacob, fellow heirs of the same promise; for he was looking for the city which has foundations,

whose architect and builder is God" (Heb. 11:8–10).
- "But as it is, they desire a better country, that is, a heavenly one. Therefore God is not ashamed to be called their God; for He has prepared a city for them" (Heb. 11:15).
- "And I looked when He broke the sixth seal, and there was a great earthquake; and the sun became black as sackcloth made of hair, and whole moon became like blood; and the stars of the sky fell to the earth, as a fig tree casts its unripe figs when shaken by a great wind. And the sky was split apart like a scroll when it is rolled up; and every mountain and island were moved out of their places" (Rev. 6:12–14).
- "And the seventh angel poured out his bowl upon the air; and a loud voice came out of the temple from the throne, saying, 'It is done.' And there were flashes of lightning and sounds and peals of thunder; and there was a great earthquake, such as there had not been since man came to be upon the earth, so great an earthquake was it, and so mighty. And the great city was split into three parts, and the cities of the nations fell. And Babylon the great was remembered before God, to give her the cup of the wine of His fierce wrath. And every island fled away, and the mountains were not found. And huge hailstones, about one hundred pounds each, came down from heaven upon men; and men blasphemed God because of the plague of the hail, because its plague was extremely severe" (Rev. 16:21).

These last two passages are from the final book of the Bible, Revelation, the only place where the Scriptures provide some specificity, albeit somewhat cryptically, concerning the final sequence of events as the Earth and all of humanity find themselves engulfed in global calamities of ever-increasing severity, including phenomena that are eerily similar to some of the specific predictions now being made by the computer model projections of climate change for the present century. So without further delay, let's tackle this challenging text.

John's Revelation (Apocalypse)

Before I begin talking about Revelation, I acknowledge that many Christians believe it to be a "difficult" book to be read only in a "symbolic" or "spiritual" sense, rather than one containing literal facts about the future.

However, the book itself plainly rejects such an interpretation with its opening words: "The Revelation of Jesus Christ, which God gave Him to show to His bond-servants, the things which must shortly take place" (Rev. 1:1). Does the book use language filled with symbolic imagery? Of course it does. But this graphic language is necessary in order to convey the sweeping scale of the topic being addressed—nothing less than the end of this world and the return of our Lord. In my opinion, to deny the coming bodily return of Jesus and the literal fulfillment of everything else written in Revelation is tantamount to denying the bodily resurrection of Jesus—without which our "faith is worthless" (1 Cor. 15:17). Ours is not merely a "spiritual religion," but a faith in both the physicality of the present created universe and the physicality of the heavenly kingdom that we will enjoy in new, glorified bodies—bodies identical to the one our Lord Jesus now possesses, still scarred from the wounds we inflicted upon him.

The words "apocalypse" and "apocalyptic" have obviously taken on very scary connotations in common discourse—all because of the apparently innocent title of this final book of the Bible, which is taken from its first few words, "The Revelation of Jesus Christ" (Rev. 1:1). However, the original Greek word, αποκαλυψις, merely means "uncovering," and, for those who love Jesus, there is quite literally nothing to fear in this book—but quite the opposite actually. In the book's prologue, John writes, "Blessed is he who reads and those who hear the words of the prophecy, and heed the things which are in it" (Rev. 1:3).

Why was this book written? The prologue answers that question very succinctly, "to show to His bond-servants the things which must shortly take place" (Rev. 1:1). How was it written? In the same verse we are told, "God...sent and communicated it by His angel to His bond-servant John, who bore witness to the word of God and to the testimony of Jesus Christ, even to all that he saw" (Rev. 1:1–2).

For those who haven't read the book, what's all the fuss about? The graphic in Figure 3.2 gives some insight. In three series of increasingly dramatic visions (seven seals, seven trumpets, seven bowls of wrath), John records the series of events that find their ultimate climax in the catastrophic physical destruction of planet Earth. I find it helpful to imagine John sitting in a vast modern-day surround-sound IMAX theatre—accompanied only by the angel—all the while struggling to follow the deceptively simple instructions blasted at him over a loud speaker, "like the sound of a trumpet, saying 'Write in a book what you see, and send it to the seven churches: to Ephesus and to Smyrna and to Pergamum and to Thyatira and to Sardis and to Philadelphia and to Laodicea'" (Rev. 1:10–11).

Neither the Bible nor secular history sheds further light on whether John literally made seven copies of the text and sent them separately to each of the seven named churches, but I personally believe that John did so, as a final measure of the loving obedience to Jesus that appears to have characterized all of his long life. This would have guaranteed the successful delivery of this important text to all of humanity, while helping to overcome the obvious questions about its reliability and authenticity—questions that continue to surround it to the present hour. My humble prayer is that this present book—which highlights the startling "coincidences" between John's ancient text and the modern-day predictions of climate scientists about the fate of our planet—will help to remove any final vestiges of lingering doubt about the faithfulness of John's text to "all that he saw" (Rev. 1:2).

Getting back to the details of Figure 3.2, the individual seal, trumpet, and bowl judgments appear to refer to the same or similar events in some cases (sixth seal seems to equate to the seventh bowl). In the case of the trumpet and bowl judgments, there is an intensification, in which the trumpets typically involve a third of the planet whereas the corresponding bowl involves the entire world. For instance, the second and third trumpets involve an apparent poisoning of a *third* of the oceans and a *third* of the fresh waters, respectively, whereas the corresponding bowl judgments involve *all* of the world's oceans and *all* of the fresh waters.

How literally are we to interpret these proportions and the sequence of events? My personal view is that they reflect an initially gradual but relentlessly quickening and increasingly drastic series of environmental catastrophes that affect larger and larger portions of the Earth's surface until the planet has become a tremendously inhospitable place. In the words of Jesus, there will be "upon the earth dismay among nations, in perplexity at the roaring of the sea and the waves, men fainting in fear and the expectation of things which are coming upon the world" (Luke 21:25–26).

Now we'll examine some of the trumpet and bowl judgments in greater detail to highlight what appear to be very close parallels between these Revelation passages and the predictions of modern science about the near-term future of our planet.

Trumpet Number One: "Hail, Fire, and Blood Thrown to the Earth"

The sounding of the first trumpet results in rampant wildfires, in which "a third of the trees were burned up" (Rev. 8:7). As we will see in Chapter 5 of this book, there is already a pattern of increasing forest fires in the western United States, and this is projected to increase in intensity as global warming continues.

Bowl Number One: "A Loathsome and Malignant Sore"

The first bowl judgment is not environmental in nature, but instead refers to a worldwide pandemic of human disease (apparently a repeat of the fifth trumpet judgment, the first of the three woes). Further global warming is expected to significantly increase the risk of such pandemics.

Trumpet and Bowl Number Two: "Blood Like that of a Dead Man"

As the second bowl of wrath is poured into the sea "it became blood like that of a dead man; and every living thing in the sea died" (Rev. 16:3). As a scientist, I ask the question, what is the first thing that happens to the blood of a dead man? As the lungs stop moving and are no longer able to expel carbon dioxide, the content of this dissolved gas increases in the blood of the recently deceased, and the pH drops as carbonic acid is formed. As mentioned in Chapter 2, this is precisely what climate scientists are now observing in our oceans as a result of the increasing combustion of fossil fuels: an increase in the level of carbon dioxide and a lowering of pH, just as in a dead man. Is this mere coincidence? I think not.

Now it is obvious that John could not have possibly understood these chemistry terms, but I believe this makes the "coincidence" all that much more humbling. For it reveals the magnificent extent of God's ability to preserve these solemn words of specific warning about the coming calamity through the many centuries. The words are there for modern man to heed, if only we would listen.

Trumpet and Bowl Number Three: "The Rivers and Springs Became Blood"

In an act recalling the first of the plagues in Egypt, the third bowl of wrath is poured "into the rivers and the springs of waters; and they became blood" (Rev. 16:4). In this particular portion of the book, there is no mention of the harm this causes, but the corresponding trumpet judgment mentions that "many men died from the waters, because they were made bitter" (Rev. 8:11). This is entirely consistent with the impact of the original plague in Egypt, when "the fish that were in the Nile died, and the Nile became foul, so that the Egyptians could not drink water from the Nile" (Exod. 7:21).

But is this particular phenomenon one of the global impacts related to man–made global warming? The linkage seems far less clear. The closest direct association is the phenomenon of acid rain, which has been a focus of environmental regulations in the United States and elsewhere since the 1980s. However, I personally feel that this is not what the passage refers to. I believe the clue is in the trumpet judgment passage, which identifies this plague with a star "called Wormwood" (Rev. 8:11). As shown in Figure 3.2, the word "Wormwood" is a direct translation from the Ukrainian language for the name of the city in northern Ukraine, "Chernobyl," which the reader will immediately recognize as the site of the world's largest known nuclear contamination incident. I acknowledge this may seem like a stretch to some, but it seems entirely possible to me that this third bowl judgment refers to a worldwide contamination of all fresh water supplies by nuclear fallout, most likely due to a future exchange of nuclear or radiological weapons.

Trumpet Number Four: "Third of the Sun Darkened"

The fourth trumpet (Rev. 8:12) apparently finds its scientific parallel in what is known as "global dimming," whereby incoming solar radiation is blocked by the excess aerosols and dusts that man has been increasingly producing. According to the latest computer modeling, the cooling impact of these aerosols was overcome during the late 1960s by the much stronger warming impact of greenhouse gases. Unlike greenhouse gases, which persist in the atmosphere, aerosols and dusts eventually settle out, leading to our next bowl judgment.

Bowl Number Four: "Men Were Scorched with Fierce Heat"

The fourth bowl is poured out "upon the sun; and it was given to it to scorch men with fire. And men were scorched with fierce heat; and they blasphemed the name of God who has the power over these plagues; and they did not repent, so as to give Him glory" (Rev. 16:8–9). It hardly seems necessary to point out the direct parallel between this event and global warming, but I guess God has "thrown it in" to help convince the most skeptical reader.

This particular judgment is unique in that it has no corresponding mention in the trumpet judgments. There is some logic to this, since it clearly refers to *global* warming, and there is no known physical mechanism for added heat to remain only in a certain portion of the planet (due to rapid atmospheric mixing). This is unlike the previous two phenomena, which referred to forms of water contamination that could remain reasonably localized for a significant period of time. Again, could John have been aware of these distinctions as he wrote them down? Of course not, but for me this only serves to raise the level of correspondence even further beyond mere "coincidence," as that term is commonly interpreted.

Bowl Number Five: "His Kingdom Became Darkened"

The next bowl of wrath appears to be related to some form of power outage throughout the civilized world the natural cause of which is not identified but is characterized in the following way, "And the fifth angel poured out his bowl upon the throne of the beast; and his kingdom became darkened; and they gnawed their tongues because of pain, and they blasphemed the God of heaven because of their pains and their sores; and they did not repent of their deeds" (Rev. 16:10–11). We can only imagine what life will be like in a significantly warmer world without the modern niceties of air conditioning, refrigerated food, etc. But widespread gnawing of tongues seems entirely possible.

Bowls Six and Seven: Armageddon and **The** Big One

Neither of these final two bowl judgments is related to climate change or global warming. Bowl number six echoes the sixth trumpet and the war-related passages of the second and fourth seals, as it speaks of armies coming upon the land of Israel from far away in the east, across the Euphrates, and eventually reaching the famous military plain of Jezreel. This plain stretches between the archeological village of Megiddo on the south and Jesus' hometown of Nazareth, on a bluff that looks down across this fertile valley from the north. The Bible has no details about the nature of the actual fighting of this battle, probably because the outcome of the human-induced carnage is rendered largely irrelevant by the great earthquake that interrupts it (bowl seven and the sixth seal).

The scale of global damage caused by the quake (**The** Big One) seems to exceed John's ability to write, but it would not be unreasonable to assume that the gravitational forces caused by the near-Earth presence of an object the size described later in the book would be sufficient to have such effects. This celestial object (named "New Jerusalem") is described by John as "coming down out of heaven from God" (Rev. 21:10). It is apparently a cube or a sphere, being described as having the following dimensions "fifteen hundred miles; its length and width and height are equal" (Rev. 21:16). I presume this one object is at once both the "new Earth and new heavens" promised so many times earlier in the Bible, for it is said that "the city has no need of the sun or of the moon to shine upon it, for the glory of God has illumined it, and its lamp is the Lamb" (Rev. 21:23). So I assume we'll all stay inside, but we'll have plenty of room. If there are ten billion of us that would give us about 1,000 acres per person, with either one thousand or two hundred foot ceilings, for a cube or sphere, respectively! For those flown directly into this city from the just-destroyed Earth, I'm sure it will seem pretty nice!

For the rest of humanity, I'm not sure which will be worse—the then-uninhabitable Earth or the coming "lake of fire" (Rev. 20:14). Unfortunately, they won't get to choose. For time will have run out. As for the earth itself, the words of Job will have found their ultimate fulfillment: "The LORD gave and the LORD has taken away" (Job 1:21).

96   REAPING THE REAL WHIRLWIND

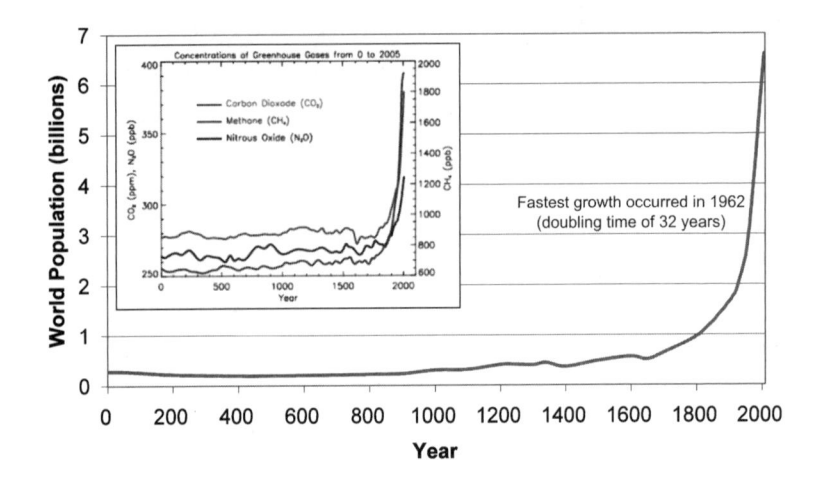

FIGURE 3.1 Median of world population estimates from the time of Christ to the present. The inset shows the growth of atmospheric greenhouse gases over the same period (sources: United Nations for population estimates and IPCC for greenhouse gas concentrations).

"The time has come ... to destroy those who destroy the earth" -- Rev. 11:18

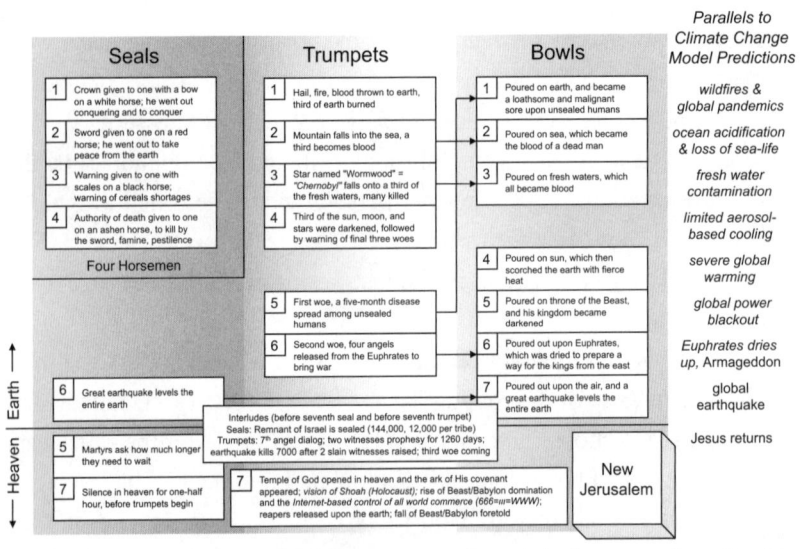

Seals	Trumpets	Bowls	Parallels to Climate Change Model Predictions
1. Crown given to one with a bow on a white horse; he went out conquering and to conquer	1. Hail, fire, blood thrown to earth, third of earth burned	1. Poured on earth, and became a loathsome and malignant sore upon unsealed humans	wildfires & global pandemics
2. Sword given to one on a red horse; he went out to take peace from the earth	2. Mountain falls into the sea, a third becomes blood	2. Poured on sea, which became the blood of a dead man	ocean acidification & loss of sea-life
3. Warning given to one with scales on a black horse; warning of cereals shortages	3. Star named "Wormwood" = "Chernobyl" falls onto a third of the fresh waters, many killed	3. Poured on fresh waters, which all became blood	fresh water contamination
4. Authority of death given to one on an ashen horse, to kill by the sword, famine, pestilence	4. Third of the sun, moon, and stars were darkened, followed by warning of final three woes		limited aerosol-based cooling
Four Horsemen		4. Poured on sun, which then scorched the earth with fierce heat	severe global warming
	5. First woe, a five-month disease spread among unsealed humans	5. Poured on throne of the Beast, and his kingdom became darkened	global power blackout
	6. Second woe, four angels released from the Euphrates to bring war	6. Poured out upon Euphrates, which was dried to prepare a way for the kings from the east	Euphrates dries up, Armageddon
6. Great earthquake levels the entire earth		7. Poured out upon the air, and a great earthquake levels the entire earth	global earthquake
	Interludes (before seventh seal and before seventh trumpet) Seals: Remnant of Israel is sealed (144,000, 12,000 per tribe) Trumpets: 7th angel dialog; two witnesses prophesy for 1260 days; earthquake kills 7000 after 2 slain witnesses raised; third woe coming		Jesus returns
5. Martyrs ask how much longer they need to wait		New Jerusalem	
7. Silence in heaven for one-half hour, before trumpets begin	7. Temple of God opened in heaven and the ark of His covenant appeared; vision of Shoah (Holocaust); rise of Beast/Babylon domination and the Internet-based control of all world commerce (666=ωωω=WWW); reapers released upon the earth; fall of Beast/Babylon foretold		

↑ Earth / ↓ Heaven

FIGURE 3.2 Summary of the catastrophic global events described in John's Revelation (Apocalypse), the final book of the Bible. Items in italics represent the interpretation of these passages as presented in this book.

Chapter Four

Predicted Climate of the Twenty-First Century

ONE KEY QUESTION that was not answered in the previous chapter seems a very legitimate one to ask: "When will all these things happen?" The reason that I was silent on this timing question is that Jesus refused to answer it. Even after the Resurrection, He was very careful to tell the disciples, "It is not for you to know times or epochs which the Father has fixed by His own authority" (Acts 1:7). Much earlier in His ministry He had already explained, "of that day and hour no one knows, not even the angels of heaven, nor the Son, but the Father alone" (Matt. 24:36).

IPCC scientists, however, are far less circumspect than our Lord. In each of the voluminous reports, they have now been dutifully producing every five to seven years, they have presented a series of very specific predictions about the future time course for numerous climate-related parameters including greenhouse gas concentrations, temperatures, precipitation patterns, and sea levels to name just a few. In this chapter I will give a simple narrative summary of these key IPCC climate predictions for the remainder of the twenty-first century, as described in their most recent (2007) set of reports.

Within both this chapter and elsewhere in this book, you will see that I question whether the IPCC modeling projections are accurate in terms of timing. I believe the data would suggest that global warming is actually proceeding considerably faster than current IPCC forecasts would suggest, probably because they have underestimated the strength of an important warming feedback process. However, I believe it is still instructive to review the IPCC forecasts in order to understand the types of changes that might be expected to take place with further warming, especially because it seems likely the predicted changes and impacts will occur even sooner than they suggest. For the reader seeking greater detail about IPCC predictions of weather conditions for the remainder of the century, I would direct you to any of the many references given in Appendix 1. I will conclude this chapter by presenting my own simple empirical analysis of the available temperature data for the land surfaces of the Northern Hemisphere. These data are available online from the US National Weather Service at http://www.ncdc.noaa.gov/oa/climate/research/anomalies/anomalies.html. I believe the observed data plainly reveal a global warming trajectory that is accelerating about twice as fast as the IPCC models suggest. The clear implication is that the current IPCC models are actually too sluggish and that temperatures will actually heat up faster than what the United Nations scientists have told us. But first, let's see what the world had to say about the latest IPCC predictions.

COMMENTS ON THE 2007 IPCC PREDICTIONS

In February 2007, when the most recent set of IPCC predictions began to come out, there was much fanfare in the worldwide media and a flurry of mainly well-orchestrated comments from numerous scientists and politicians from around the world, as summarized in a special issue of the British scientific journal, Nature (although one or two contrary voices were recorded, as can be seen below):

- *"This may be remembered as the day the question mark was removed from whether human activity has anything to do with climate change."*
  —Achim Steiner, head of the United Nations Environment Program
- *"Now is not the time for half measures. It is the time for a revolution."*
  —French (now former) president Jacques Chirac
- *"The question is, what can we do now? There's very little we can do about arresting the process."*
  —Anote Tone, president of the Pacific island nation of Kiribati
- *"This should compel all of us towards action rather than the paralysis of fear."*
  —Martin Ress, president of the United Kingdom's Royal Society

- *"Now it's time for us—the policymakers—to do our jobs."*
  —Bart Gordon, Democratic Congressman from Tennessee and chair of the US House Committee on Science
- *"This is a group of climate experts attempting to reach a scientific consensus. It doesn't commit governments to any course of action."*
  —Pradipro Ghosh, senior official at India's Ministry of Environment and Forests
- *"For sure, humans cause global warming!"*
  —Headline from China's Xinhua news agency
- *"Let's be realistic. You can only run power stations in a modern Western economy on fossil fuel, or, in time, nuclear power."*
  —Australian (now former) prime minister John Howard, whose country finally ratified the Kyoto protocol in December 2007, per a campaign pledge of Howard's successor
- *"Those who continue to ignore the threat will be doing the greatest disservice imaginable to current and future generations."*
  —Marthinus Van Schalkwyk, environmental affairs minister for South Africa

IPCC Summary of Physical Science Basis

This IPCC summary document is intended to inform policymakers about the high level conclusions of the scientists without forcing the lawyer-politicians to read the "thick documents." It begins with a discussion of the human and natural drivers of climate change, pointing out that global atmospheric concentrations of the three most important human-derived greenhouse gases now far exceed any levels of at least the past one million years. These gases include carbon dioxide, methane, and nitrous oxide. The increases of carbon dioxide are due primarily to fossil fuel use and land use change, while those of methane and nitrous oxide are primarily due to agriculture.

Throughout the reports, IPCC states the warming impact of greenhouse gases in terms of "radiative forcing," which refers to the amount of energy per unit area (watts per square meter, abbreviated W m-2) that the gases now trap in the atmosphere measured relative to the condition of the atmosphere before the year 1750. The report states that the understanding of net anthropogenic warming and cooling influences has improved since the 2001 IPCC report, leading to a very high confidence (>90% likelihood) that the global average net effect of human activities since 1750 has been one of warming with a radiative forcing of 1.6 W m-2 (with a 90% certainty interval ranging from 0.6 to 2.4 W m-2). The report states that current warming is unequivocal, with eleven of the last twelve years (1995–2006) ranking among the twelve warmest years in the instrument record (since 1850).

The summary states that most of the observed warming is very likely (>90% confidence) due to increased greenhouse gases and that continued warming of about 0.2°C (0.36°F) per decade in global mean temperatures is currently expected according to IPCC modeling. Significantly, the report states that warming and sea level rise will continue for centuries, even if greenhouse gas concentrations could be miraculously stabilized at current levels, due to numerous feedback processes and slow vertical mixing of the oceans.

A particular telling figure is on page 18 of the report (Figure SPM–4), which shows the retrospective twentieth century modeling runs with and without human-derived forcing (greenhouse gases). The red bars (with forcing) show far better agreement with the measured temperatures for all regions and for the global mean. However, it is also clear in these figures that the observed land temperature warming is at the upper end of the 90% confidence band in the modeling, which (as pointed out earlier in this chapter) suggests that the models are too "sluggish" in their predictions of current and future warming trends. This fact implies that the models are underestimating the strength of a feedback process, most likely the warming impact of increased atmospheric water vapor.

The IPCC states that there have been significant advances in the reliability of the regional climate change projections. Some of the major conclusions include the following: The amount of warming over many land areas is greater than the global mean due to less water available to provide evaporative cooling and the greater heat capacity of water than land. According to the fourth assessment report, warming is very likely over all land masses. Precipitation effects are much more difficult to predict with the models, but the trend is for a higher proportion of the rainfall to come in the form of convective thunderstorms. There is also a general trend for a reduction of rainfall in the subtropics and an increase in higher latitudes and parts of the tropics. These rainfall predictions for tropical regions are very sensitive to the amount of tropical storm activity that accompanies global warming, and the various models disagree on the overall impact.

IPCC Scenarios

Since the IPCC modeling runs attempt to predict future human impacts on climate, it is necessary for them to make assumptions concerning trends in global and regional population growth, "decarbonization" of transportation fuels and energy production, land use patterns, control of aerosols and other forms of air pollution, and changes in agriculture. This has been accomplished by the development of a family of six scenario groups, each of which has made slightly different assumptions about future trends in human development. They are briefly described below.

Scenario A1: The A1 scenario family describes a future of very rapid economic growth, global population that peaks in mid-century (2050) and declines thereafter, and the rapid introduction of new and more efficient technologies. The three A1 scenarios are distinguished by their technological emphasis: fossil fuel intensive (A1FI), non-fossil fuel energy sources (A1T), or a balance across all sources (A1B) (where balanced is defined as not relying too heavily on one particular energy source).

Scenario A2: The A2 scenario describes a very heterogeneous world. Economic development is primarily regionally oriented and per capita economic growth and technological change are more fragmented and slower than in other storylines.

Scenario B1: The B1 scenario describes a convergent world with the same global population as in the A1 storyline, but with rapid change in economic structures toward a service and information economy, with reductions in material intensity and the introduction of clean and resource efficient technologies.

Scenario B2: The B2 scenario describes a world in which the emphasis is on local solutions to economic, social, and environmental sustainability. It is a world with continuously increasing global population, intermediate levels of economic development, and less rapid and more diverse technological changes than in the B1 and A1 storylines.

The scenarios are "conservative" in the sense that none of them assume successful implementation of United Nations climate initiatives, such as the notorious (at least here in the United States!) Kyoto Protocol. Not surprisingly, overall model predictions end up being quite sensitive to the choice of scenario, because each has very different outcomes in terms of global human population and the pattern of growth of global greenhouse gas emissions. In the discussion that follows, I will focus on the median predictions of one of the IPCC modeling scenarios (denoted scenario A1B), but ranges will be given when appropriate. Of the six scenarios, A1B predicts warming at the third fastest rate. The warming rate order of the six scenarios, from fastest rate of warming to the slowest, is as follows: A1F1, A2, A1B, B2, A1T, B1.

IPCC Projections for the Twenty-First Century

One of the many advances in modeling technology that IPCC highlights in the new reports is that there is a much larger number of simulations from a broader suite of models. When combined with the fact that so much more observational data are available, the result should be that the certainty of predictions about future climate trends has been significantly enhanced, with a much more reliable means of characterizing uncertainty in the model predictions. However, I don't believe the IPCC scientists have paid very much attention yet to the previously identified "under prediction" of the current warming trajectory. Perhaps the observations highlighted in this book will influence IPCC to take a closer look at this. But I confess I am not optimistic!

The IPCC displays results for a wide range of assumptions concerning future trends in human population and the degree to which changes are made in the utilization of fossil fuels. The scenarios include a very highly idealized case in which both greenhouse gas and aerosol emissions are "miraculously" held constant at the year 2000 levels. This is clearly a highly idealistic and unrealistic case, but it does give an "idealized best case" against which to judge all of the more reasonable scenarios.

The typical rate of warming expected for the early part of the twenty-first century is for warming of 0.2°C (0.36°F) per decade as a global mean.

For the "idealized best case" of year 2000 emission levels held constant, the rate of warming would be about half that amount. A sensitivity analysis was conducted to investigate conditions during the period from present through the year 2030, and IPCC scientists concluded that the rate of warming over all land masses is relatively insensitive to the choice of scenario and will be at least twice as fast as the warming observed during the twentieth century.

In the detailed numeric discussion that follows, the "datum," as defined by IPCC for discussing the various warming predictions, is the observed mean temperature over the final twenty years of the twentieth century: 1980–2000. The IPCC projections for the various energy utilization scenarios begin to diverge rapidly through the latter half of the twenty-first century. For the A1F1 scenarios, corresponding roughly to a "business as usual" set of assumptions, the expected amount of global warming during the twenty-first century is about 4.0°C (7.2°F), with a 90% upper bound estimate of about 6.5°C (11.7°F) for the global mean. The warming over land areas would be greater.

By contrast, the B1 scenario, which is the "greenest" in terms of the assumptions about humanity's adoption of less carbon-intensive energy sources, results in a predicted degree of warming of only 1.8°C (3.2°F) by the end of the century. The overall range of warming predictions for the twenty-first century that are given in the 2007 report is very similar to the range provided in the 2001 IPCC report: 1.4–5.8°C (2.5–10.4°F), though the authors of the 2007 report are quick to point out that the numbers are not truly directly comparable, due to improvements in the way that the uncertainties are being expressed.

One factor that causes upward curvature in the model predictions under the high carbon dioxide emission scenarios is that the models assume a certain degree of "climate–carbon feedback," in which the ability of both the ocean and land surfaces to absorb the carbon decreases as the temperature increases. Obviously, if the modelers have gotten the sensitivity of this warming feedback wrong, the predictions could be way off—in either direction. Looking at the recent data, it seems reasonable to conclude that the warming feedback strength is not quite right due to the trend for the models to "undershoot" the current warming trajectory.

For each set of warming scenario predictions, the IPCC team has reported a corresponding range of projected sea level rise estimates for the end of the century. Across all six scenario families, the overall range of sea level rise is 0.18–0.59 m (7–23 inches). The IPCC scientists note that higher numbers cannot be excluded, primarily due to the possibility of "rapid dynamical changes in ice flow," but the overall range of projected sea level rise is actually a bit less than what was presented in the 2001 IPCC reports.

More troubling, however, than the projected sea level rise, is the observation that the continuing uptake of carbon dioxide by the oceans is actually lowering the pH of surface seawater, already by 0.1 pH units (a 30% increase in hydrogen ion concentrations). The range of future ocean acidification quoted in the IPCC report is an additional 0.14 to 0.35 pH units over the twenty-first century, but greater drops in seawater pH have been projected by other scientists. As mentioned in Chapter 2, the reason this is so troubling is that the increases in ocean acidity dramatically increases the solubility of calcium carbonate, the very material that forms the shells of many marine organisms including coral, mollusks, and a number of microscopic organisms. The effect of this pH drop have only begun to be studied, but the results collected thus far suggest that entire marine food webs, including key fisheries used as important food sources by several nations, are at risk.

The geographic patterns of warming are independent of the carbon utilization assumptions in the modeling—it's just the timing that changes. Warming is projected to be fastest over those portions of the globe dominated by land masses, the poleward portions of the Northern Hemisphere, where snow-cover, glaciers, and permafrost are all expected to contract quite rapidly—as will sea ice, and the ice sheets of Greenland and the Antarctic. In several of the IPCC scenarios, sea ice almost completely disappears during late summers in the Arctic by the end of the twenty-first century. On this particular topic, it is worth noting that recent satellite observations show far more summertime loss of the Arctic ice cap than predicted by most IPCC models—with some scientists now forecasting complete summertime loss of the Arctic within only a few years.

Not surprisingly, with all of the additional heat energy pumped into the atmospheric system, the models predict major increases in extreme weather such as heat waves, severe thunderstorms, wind events, and flash flooding. Storm tracks in the sub-tropics and higher latitudes are projected to move poleward in the modeling, which will contribute to precipitation tending to increase near the poles and current drought prone areas to expand toward the poles.

As for tropical cyclones (hurricanes and typhoons) the models diverge greatly in their predictions. A slim majority of the models predict an increase in the number of intense (category 5) storms, in direct relationship to the predicted increase in summertime sea surface temperatures. However, some models project a decrease. Confounding this entire discussion of the model predictions for such storms is the fact that the apparent increase in the number of very intense tropical cyclones since 1970 is considerably more than the model runs for the same period of time. In short, the models seem to be reluctantly leaning in the direction of more storms, but this may be another area where the model response to increased heat is actually too sluggish compared against the reality of what is actually being observed.

As for the ocean circulation, the most critical current is the Meridional Overturning Circulation (MOC) of the North Atlantic. This current is responsible for bringing relatively mild weather to northern Europe, without which agriculture there would be very difficult to maintain. It is also the part of the "oceanic conveyor belt" that was apparently completely shutdown about 8,200 years ago, as a result of a sudden entry of fresh water into Hudson Bay at the end of the most recent period of glaciation. Although the MOC circulation is projected to slow somewhat, the deceleration is not expected to be enough to slow the warming trend in northern Europe, and the IPCC modelers are confident that it will not be completely shut down during the twenty-first century. However, behavior beyond that point is very uncertain with some models projecting a collapse in the twenty second century.

Near the end of the *Summary for Policymakers*, the IPCC scientists quote a number of targets for carbon emissions that would be necessary to cause atmospheric carbon dioxide levels to level off. Part of the challenge in such calculations is that the strength of the climate–carbon feedback process is still not known with certainty. Nevertheless, the scientists provide estimated targets for cumulative carbon releases that would result in stabilized carbon dioxide concentrations of 450 and 1,000 PPM. Subsequent to the release of the 2007 IPCC report, Dr. Jim Hansen (NASA) has recently (April 2008) presented an analysis suggesting that *lowering* carbon dioxide to a level of 350 PPM is the only strategy that avoids "catastrophic consequences." Hansen's formula for accomplishing such a reduction includes: (1) phasing out of all coal use (without carbon capture and storage), (2) elaborate new schemes for trapping of carbon in our management of agricultural and forestry systems, and (3) no further exploitation of new oil reserves.

Hansen's target of 350 PPM, however sound, seems unlikely to become public policy. Even the much more modest IPCC target of 450 PPM would require keeping cumulative carbon dioxide emissions for the century at 1,800 gigatons. Given that current annual emissions are already well over twenty-five gigatons per year and growing, this would appear to be an excessively challenging target. According to current model estimates of the climate–carbon feedback, stabilizing at 1,100 PPM would require keeping total emissions for the century at no greater than 4,000 gigatons, which seems within reach without too much change in current human behavior. However, that would also correspond to a quadrupling of the incremental radiative forcing due to man-made carbon dioxide, and the world is already warming faster than predicted with the current concentration of this gas: about 385 PPM at the end of 2007.

The IPCC scientists also present much longer term assessments of what might take place after the year 2100. Even if mankind is somehow able to stop the increase in greenhouse gas concentrations sometime during the century (and no realistic scenario envisions that possibility), the Earth would continue to warm for more than 1,000 years, due to the time scales required for the removal of this gas from the atmosphere.

IPCC Regional Predictions for the Twenty-First Century

When just one chapter from an IPCC assessment report is printed, one appreciates the practical utility of each "Summary for Policymakers" that the IPCC prepares. One example is the eleventh chapter of the following tome: *Climate Change 2007: The Physical Science Basis: Contribution of Working Group I to the Fourth Assessment Report of the Intergovernmental Panel on Climate Change*. The chapter is devoted to regional projections and comprises pages 847 through 940 of that particular report. It goes into graphic detail (literally) on what is described as the "increasingly reliable regional climate projections" that are now available due to advances in modeling and understanding of the physical processes in the climate system.

A key theme of the chapter on regional projections is that warming over many land areas is much greater than global mean warming due to less water availability for evaporative cooling and the smaller thermal inertia of land as compared to the oceans. Anyone who has spent a "summer day" shivering on the peninsula of San Francisco can appreciate the physical reality of that fact. Another theme is that warming will increase the spatial variability of precipitation throughout the world, contributing to a reduction in rainfall for the subtropics and an increase at both higher latitudes and in some parts of the tropics.

Subtropical high pressure systems are projected to move poleward with the predicted impact of major reductions in precipitation at the poleward edges of the subtropics. Monsoonal circulations are predicted to become enhanced, leading to higher tropical moisture, but the models have less of a consensus on this trend. The IPCC is very careful not to make any firm conclusions about trends in tropical cyclones (such as Atlantic hurricanes), though some modeling results suggest an increased number of category 5 storms. The bulk of the IPCC regional discussion is devoted to specific maps of particular areas, region by region, the highlights of which are presented here.

Africa

As with all land masses, IPCC predicts the warming of the continent will very likely be greater than the global mean, with precipitation changes distributed unevenly as on the other continents. Most of northern Africa, and especially the portions near the Mediterranean, are expected to become drier, which will facilitate a continuing expansion of the Sahara desert. Similarly, the southern and especially southwestern portions of Africa, much of which is already arid, will become drier. The only regions likely to experience increases in rainfall are in east Africa and the equatorial regions of the west coast of the continent.

The report acknowledges that there are technical limitations to the predictive capacities of the current computer models when it comes to predicting precipitation patterns at the regional scale. The result of this unsettled science is that differing assumptions are being made by the model authors, based on their personal guesses of which parameters will become dominant in determining future precipitation patterns. The only objective way to determine the accuracy of the various models is to check how well they have predicted rainfall patterns for the twentieth century, and the generally poor ability of today's models to accomplish this for Africa is good evidence that all of the rainfall predictions should be viewed as tentative, at best.

IPCC suggests that part of the reason for poor prediction of rainfall in Africa may be due to a failure to include important physical processes in the models. Two potential feedback processes not in the current models include vegetation feedback and feedback from dust aerosol production. In addition, the models fail to account for future land surface modification, such as might be caused by either extensive deforestation or reforestation efforts.

Africa's rainfall patterns are dominated by seasonal movement of the tropical rain belt. In addition, both the northern and the southern edges of the continent are far enough from the equator to be impacted by midlatitude fronts, which are generally expected to move poleward as global warming intensifies. Another key factor influencing Madagascar and southeastern Africa is the possible intensification of tropical cyclone activity.

Europe and the Mediterranean

According to IPCC, Europe will warm at a rate faster than the global mean temperature rise, with northern Europe warming even more dramatically during the winter and the Mediterranean areas warming most in the summer. In a fashion parallel to that seen in the temperature patterns, annual precipitation amounts will increase in northern Europe and decrease in the Mediterranean. In central Europe, the effects on precipitation are predicted to counterbalance on an annual basis, with winter increases offset by summertime decreases. Summertime droughts will become increasingly likely in both central and southern Europe. A detail in the models that is not generally presented for other continents is that northern Europe is expected to experience an increase in the frequency of extreme wind events. Finally, and not unexpectedly, the duration and intensity of snowy periods are both expected to lessen dramatically as warming proceeds.

Much of central Europe experienced an intense and deadly heat wave during the summer of 2003, resulting in over 30,000 deaths. Other recent significant weather events have included the severe flooding of central Europe in August 2002 and several decades of much warmer winters. These events are believed by IPCC to be correlated with changes in the North Atlantic Oscillation (NAO), a cousin to the notorious El Nino and La Nina that we hear so much about here in the United States.

The United Kingdom and other areas of northwestern Europe are unusually mild, in relation to their far northern latitude, due to the northward conveyance of heat by the Gulf Stream. This flow of warm water subsequently sinks in the North Atlantic, and is now believed to be one of the most critical portions of the global oceanic conveyor belt system (the MOC, which was introduced earlier in this chapter). Most climate models predict the MOC will slow during the twenty-first century, which should somewhat retard the rate of warming in northern Europe, but there is no expectation that this could actually reverse the overall warming trend for that portion of Europe.

There are several climate feedback processes that are particularly important in helping to determine European weather. For instance, the extent and duration of snow cover helps to cool northern Europe during the winter by reflecting more sunlight. However, with less snow falling and staying on the ground, more dark surfaces remain uncovered, allowing more sunlight to be absorbed and thereby warming the continent even faster. As for central and southern Europe, which only rarely experience snowfall, the key feedback is in the summer months, when the drying out of soils serves to accentuate heat waves through the reduction of evaporative cooling and other feedback effects, such as a reduction in cloud formation.

Asia

IPCC predicts the rate of Asian warming will be greatest in Siberia, Tibet, and the central desert portions of the continent. With the exception of Southeast Asia, the remainder of Asia will experience temperature increases somewhat above the global mean. The rate of warming in Southeast Asia will be very close to the overall global mean due to the proximity of the oceans, and their accompanying high thermal inertia. Summer heat waves and droughts in eastern Asia are very likely to increase in duration, frequency, and intensity. Precipitation is expected to increase most noticeably during winter months in both northern Asia and Tibet, but general annual increases are expected throughout most of Asia. The only exception to the general precipitation increase will be in central Asia, where the current desert area is likely to increase in spatial extent. Extreme rainfall events in eastern and Southeast Asia, especially those due to tropical cyclones, will increase in frequency and severity, although monsoonal flows may weaken.

As the climate predictions for Asia are presented, the IPCC scientists are careful to note that the highly complex topography and marine influences presented by the geography of Asia are major challenges for today's climate models and are actually largely ignored. Therefore all of the general trends presented in the report become suspect for particular locations within the larger modeled regions.

Monsoons are a major climate phenomenon throughout much of Asia, especially the most heavily inhabited regions near the coasts. Global warming simulations typically predict a weakening of monsoonal flows, particularly in the overall atmospheric flow velocity from oceans into the land regions. However, these moderating influences may be more than offset by the strengthening impact of higher moisture content caused by higher temperature. The current consensus of the modelers, therefore, is for increases in overall precipitation—especially as the impact of possibly more frequent intense tropical storms is factored into the analysis.

The Tibetan plateau is particularly challenging for the climate models, due to numerous factors. Observational data are sparse, and there are many competing effects from adjacent regions. To the west, the general warming and drying of the Mediterranean will have an effect. To the far north, the Arctic will be warming and becoming wetter, but the intervening desert will be expanding. Finally, the extremely high altitude and excessively rugged and complicated topography of the Himalayas present obvious challenges for the climate models, which are largely calibrated to deal with flatter, near sea level geographies. But this much is certain: the current extent of glaciation on the Tibetan plateau will decline quite dramatically—a fact that will influence the availability of fresh water in downstream communities, mainly in China.

North America

The pattern of warming predicted by IPCC for North America is similar to that of Europe, with the northern regions experiencing the most warming in the winter months and the Southwest seeing the most warming in the summer. As in Europe, this is reflected in the precipitation pattern, which shows the northern regions becoming wetter and likely decreases in the American Southwest. Southern Canada will experience wetter winters and springs and drier summers. The extent of snowfall, both in duration and depth, is likely to decrease throughout most of the continent, with the possible exception of extreme northern Canada.

The greatest source of uncertainty associated with the climate model predictions for North America are associated with the potential lessening of ocean circulation patterns (like the MOC) that may be strong enough to actually cause localized cooling of New England and the Maritime Provinces. Also contributing to overall uncertainties are questions about the reliability of the forecasts for changes in Atlantic hurricanes, which are generally predicted to increase in both frequency and severity, but these storms are not actually resolved by the coarse-scale models used by IPCC in making their general regional forecasts through the end of the twenty-first century. So the tropical storm forecasts rely upon ancillary modeling studies of lesser scientific rigor.

The dominant weather feature throughout much of North America, especially the central United States, is the storm track, which is projected to move only slightly north, but be fed by a significant increase in atmospheric moisture content. This is predicted to result in a general increase in precipitation throughout most of the continent. Another important aspect of the storm track as it becomes more vigorous is that there will be an enhancement of nocturnal storms in the Plains region of the United States, especially in the summer months, meaning that an ever increasing proportion of rainfall for these areas will come in the form of thunderstorm activity. The key exception to this will be both Mexico and the southwestern portions of the United States, both of which already experience large annual moisture deficits—a trend that will intensify.

CENTRAL AND SOUTHERN AMERICA

For all but extreme southern South America, IPCC expects the rate of warming will exceed the global mean, due to the general trend for faster warming of land surfaces. The extremely narrow southern tip of the continent will warm at the same rate as the more slowly warming ocean. Annual precipitation is expected to decrease in Central America, as already mentioned for Mexico. Precipitation increases are expected during the winter in Tierra del Fuego and during the summers in Argentina. However, precipitation patterns for the remainder of South America are much less certain. There is a suggestion that the current trend of increasing annual rainfall in the Amazon River basin will continue through the century, but the models do not agree on this forecast.

For much of this entire region, tropical storms are the key feature determining annual rainfall. Other unique aspects of the region's weather include the presence of the Amazon rainforest and the rain-shadow effect of the nearly continuous Andean mountain range. Southern South America is also influenced by extratropical disturbances, analogous to those that dominate the weather of much of central North America.

AUSTRALIA, NEW ZEALAND, AND SMALL ISLANDS

IPCC predicts Australia and New Zealand will warm only slightly faster than the global mean due to the moderating influence of the surrounding oceans. Southern and southwestern Australia are expected to experience severe decreases in precipitation. Northern and northeastern Australia are impacted by monsoonal flows and tropical cyclones, so much of the predicted precipitation increases are directly related to the uncertain trends already mentioned for those two types of weather features.

The small islands of the world are expected to experience warming that is somewhat less than the global mean in all seasons. The expected sea level rise of 0.18–0.59 m (7–23 inches) by the end of the century will have varying impacts depending on local topographical features. Any changes in annual precipitation for all of the small islands are extremely uncertain, because the current global and regional models do not even attempt to represent these islands. Thus all predictions are actually for the surrounding oceans, and are almost certainly erroneous for even very small bodies of land. Relatively little effort has yet gone into developing improved estimates for these regions, which is somewhat understandable due to the small voice they have in the United Nations.

POLAR REGIONS

IPCC expects the Arctic to warm more rapidly than the global mean with an increase in annual precipitation, especially in the winter. Similar trends are predicted for the Antarctic, but the rate of change will not be as dramatic. Predictions of climate for both poles are at the extreme edge of current climate modeling capabilities. The extent and thickness of Arctic sea ice will undoubtedly continue to decrease, as is already being observed. The disappearance of the Arctic ice cap introduces an extremely important feedback into the overall rate of global warming, through both surface albedo (reflectivity) and thermal effects.

For the Antarctic, the weather feature that has received the most attention is the predicted increase in snowfall and the resultant impact on the interior ice sheet. The atmospheric circulation pattern will also be modified, according to the simulation models, which should also influence the overall rate of ice sheet accretion. This phenomenon has some limited feedback potential, but it is far less important than the predicted seasonal disappearance of the Arctic ice cap.

NEW SCIENCE NOT FULLY ADDRESSED BY IPCC

The IPCC modeling process involves a great deal of complicated coordination between scientists from around the world, extensive interaction with United Nations policymakers, and the individual governments that are its member nations—which all means that it is intrinsically slow to respond to the latest science that comes from outside its panel of direct contributors. Indeed, even scientists directly serving on the IPCC may have difficulty being heard. The net result is that an important amount of new science is necessarily excluded from each report—not due to any "conspiracy" on the part of the IPCC—but rather as a direct consequence of the sheer size of the IPCC effort. A few examples of such new science are briefly discussed here.

One example of this (Diffenbaugh et al., 2005) focuses on the effect that greenhouse gases will have on the frequency and intensity of extreme weather events and the interaction of these atmospheric processes with fine-scale climate system modifiers, such as snow-cover reflectivity and "rain shadow" (reduced down-wind precipitation) effects near mountainous regions of the western United States. These authors project substantial, spatially heterogeneous increases in both hot and wet events over the contiguous United States by the end of the twenty-first century and go on to suggest that consideration of fine-scale processes (not currently modeled by IPCC) are critical for an accurate assessment of local and regional-scale vulnerability to climate change.

Another example of new science is a fascinating paper by LeGrand et al. (2006), which describes new advances in modeling the last major abrupt climate event (about 8,200 years ago), during which time the Northern Hemisphere experienced a sudden cooling. This rapid event interrupted an overall warming pattern as Earth emerged from the most recent period of heavy glaciation. For some time scientists had been stymied in their efforts to reconcile differing lines of evidence as to what caused this sudden cooling. However, LeGrand and his coauthors argue that new modeling confirms the lead hypothesis that a catastrophic drainage of the ancient glacial Lakes Agassiz and Ojibway into the Hudson Bay were a catalyst for suddenly stopping the MOC in the North Atlantic. The new simulations produce a short period of a significantly diminished MOC current, sufficient to match the vast majority of other observations.

In another recent article (Seager et al., 2007), a group of scientists use the latest advances in regional climate modeling to conclude that a transition to a more arid climate is already underway and intensifying in the southwestern portions of North America. The most ominous of the predictions is the trend toward a growing moisture deficit throughout the region. They note that all six of the multi-year droughts in the instrumental temperature record have been linked to variations in the sea surface temperatures of the tropical Pacific Ocean and that future droughts will be longer and of greater severity, because they will be perturbing a base state much drier than any experienced recently.

In a recent news article (2007) Schiermeier attempts to give a broad and balanced view on the many parts of the underlying science that are not yet settled in the IPCC predictions. At the top of the list of unknowns are the many feedback processes in the models: carbon cycling, ocean currents, decomposing organic matter in the tundra, impacts on oceanic life, etc. The question of sea level rise is also very much in play, with the 2007 IPCC report representing a bit of a retreat when it came to predicting sea level rise, only 0.18–0.59 m (7–23 inches) by 2100, far less than in the prior reports. However, this is still being "hotly" debated within the climate modeling community. Schiermeier mentions a recent article by Stephan Rahmstorf in *Science*, published online in February 2007 and therefore much too late to influence the mammoth IPCC consensus documents. Rahmstorf still predicts a rise of 1.4 m (55 inches) by the year 2100.

A Closer Look at the Current Rate of Warming

For those intimidated by mathematics, please feel free to skip ahead to Chapter 5, but for the interested and adventurous reader, I invite you to join me for a brief but closer look at the observed rate of warming over land surfaces in the Northern Hemisphere (see Figure 4.1). These data come directly from the US National Weather Service and are available online at http://www.ncdc.noaa.gov/oa/climate/research/anomalies/anomalies.html. I have plotted the data in degrees Fahrenheit, rather than Celsius, in order to make them more meaningful for those of us in the United States. As a simple data analysis tool that is entirely reasonable for such noisy data, I have added a seven-year moving average (centered) to Figure 4.1, in order to see the overall trend a little more easily.

The striking thing that I noticed when I added this seven-year moving average to the graph is that it is smoothly nonlinear and concave upward for the past forty years. Why has this very strong warming signal suddenly appeared in the record? I'm sure the IPCC scientists would give a variety of guesses, but my personal opinion is that it is likely a result of the carbon dioxide warming effect finally becoming dominant over the mix of other man-made activities that have a net cooling effect, especially conventional air pollution due to particulate matter. The upward curvature is also consistent with positive feedback being induced by increased evaporation of water into the atmosphere with that warming itself adding to the overall greenhouse effect.

Whatever the actual cause of the emergence of this accelerating warming curve, I have found that it is fit extremely well by the following equation, which I obtained by simple least squares regression to the seven-year moving average of the observed data since 1968. It is a quadratic in terms of time:

$$T = [a(Y - 1968)^2] + [b(Y - 1968)]$$
where T is the Northern Hemisphere land surface warming
relative to the year 1968 (°F),
Y is the year (conventional Gregorian calendar),
a is 0.0008338 °F/yr^2, and
b is 0.024337 °F/yr.

As is plainly visible in Figure 4.1, this quadratic fit predicts much faster warming than the IPCC model predictions for the decade of the 2020s, consistent with the observation I have made earlier in this chapter—that the IPCC models are overly sluggish in their predictions of the current warming trajectory. To be fair, I must note that this comparison is for the land surfaces of the Northern Hemisphere only, but that is the part of the world that most of our population inhabits. I will leave the work of making such comparisons for less populated world areas to other interested scientists, but I suspect that a similar trend will hold true.

As for the simple quadratic fit that I am proposing for the current rate of warming in the Northern Hemisphere, I have taken a closer look at how well it is fitting the period since 1968 (see Figure 4.2). In my experience as a practicing environmental scientist, I find the degree of fit to be surprisingly good, and I suspect that it will continue to fit the observed warming for a considerable period of time in the absence of some tremendous intervening event—such as a massive volcanic eruption. I have been checking the US National Weather Service data on a monthly basis since I first developed this equation, and each new month's value continues to track along this curve quite nicely (see the inset in Figure 4.2), sufficient to convince me it is a useful model for the current warming trajectory—however simplistic it may appear to my IPCC brethren!

So what does this simple equation predict for the rest of the twenty-first century? The amount of further warming predicted by this equation for the year 2063 is about 8°F, by which time current greenhouse gas emission trends would bring atmospheric carbon dioxide to 556 PPM, double the preindustrial level of 278 PPM. This degree of climate sensitivity to carbon dioxide is very close to what Arrhenius first predicted over a century ago. As shown in Figure 4.3, the equation predicts a rate of temperature increase that is roughly double the midpoint of the current IPCC model predictions. Which is right? Only time will tell. But don't bet your future on the IPCC models being overly aggressive in their predictions on the rate of warming. I suspect they will continue to be very conservative for the foreseeable future—largely as a result of the complex consensus-driven political process that apparently causes them to be too cautious.

Meanwhile, things will continue to get hot down here!

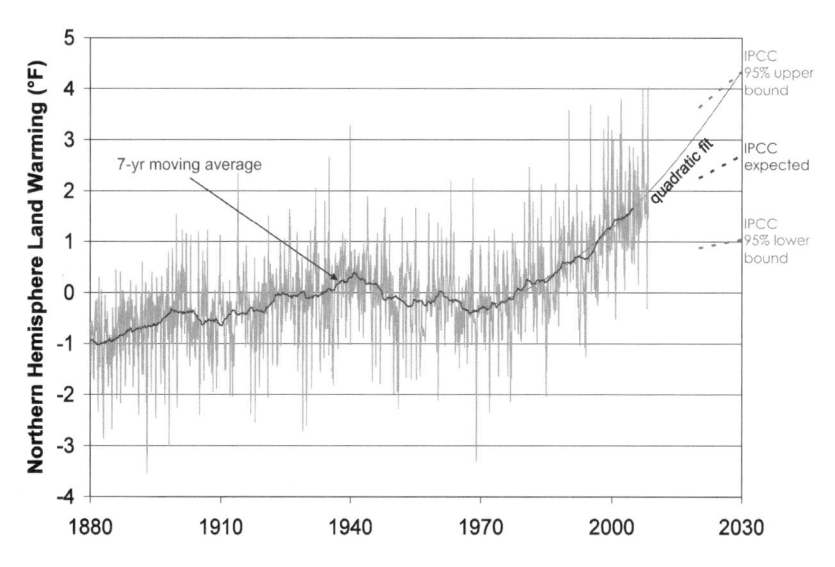

FIGURE 4.1 Observed Northern Hemisphere land surface temperatures are shown along with the seven-year moving average, a quadratic fit to this moving average from 1968 to the present time, and the median of current IPCC predictions for the warming trend in the Northern Hemisphere during the decade of the 2020s (source: NOAA).

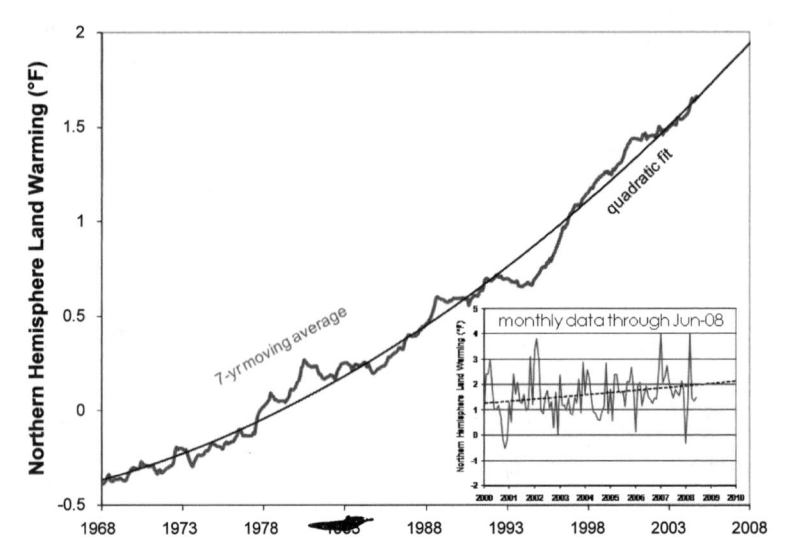

FIGURE 4.2 Observed Northern Hemisphere land surface temperatures for the past forty years in comparison with the quadratic fit. The inset at lower right shows the unsmoothed monthly values up to the most recent month available when the book went to press (June 2008).

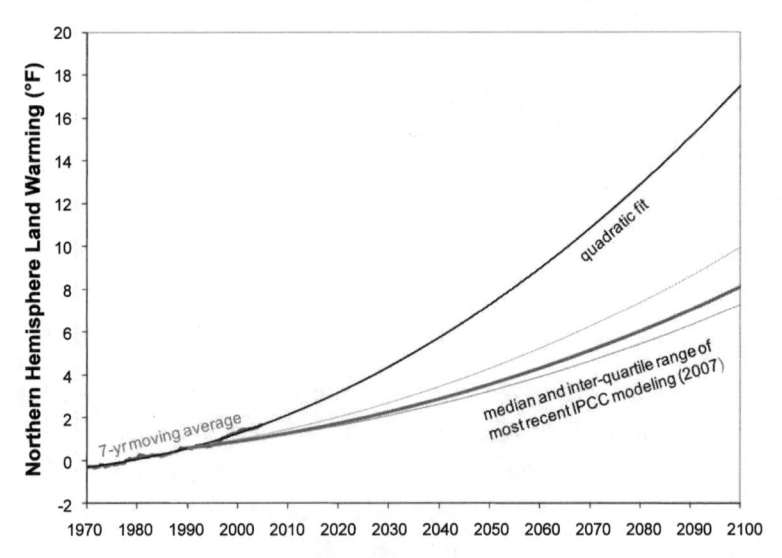

FIGURE 4.3 Comparison of current IPCC model predictions and the quadratic fit for Northern Hemisphere land surface temperatures for the remainder of the twenty-first century.

Chapter Five

PREDICTED CLIMATE CHANGE IMPACTS

IN THIS CHAPTER I WILL AGAIN rely primarily on the most recent IPCC reports to examine the impacts of climate change on natural and human systems. But I will also take some time to summarize some of the additional work that has recently been published on the related issues of ocean acidification and hypoxia (the phenomenon by which "dead zones" of reduced oxygen content occur where certain rivers discharge into the ocean). This overall discussion now begins to go far beyond simple climate science. It now becomes necessary to consider the resiliency of physical, biological, and social systems to cope with the challenges that man-made global warming is expected to bring. The summary I will present is meant only to be a brief overview of the impacts. The reader interested in more detail is encouraged to consult the IPCC source documents and the other references listed in Appendix 1.

WHAT ARE THE CURRENT IMPACTS?

The IPCC analysis of current impacts focuses on the impacts of climate change on natural and human systems, their ability to adapt to these changes, and their vulnerability to further changes. The timescale considered is the nearly forty year period since 1970. The IPCC points out that the number and quality of studies on impacts has greatly expanded since the third IPCC assessment in 2001. However, the IPCC also admits there is a lack of geographic balance in the available information, with most studies having been conducted in Europe. There is a glaring lack of information from most agricultural regions and any developing countries.

As for the natural systems, IPCC highlights the impact that warming is already having on glacial lakes, ground instability in permafrost regions, rock avalanches in mountainous areas, changes in Arctic/Antarctic ecosystems, increased and earlier peak spring runoff, warming of lakes and rivers, earlier biological responses in springtime, and poleward shifts in numerous biological species (both on land and in the oceans). With regard to human systems, IPCC states that there have already been changes in the timing of seasonal activities in agriculture, forestry, human health, and in the Arctic. The IPCC also cites increased risks to communities in mountainous regions, drought-related suffering, and increased flooding of many coastal regions.

Future Impacts

The IPCC highlights likely impacts on fresh water resources in many critical regions, threats to certain ecosystems, changes in the geographic distribution of agricultural and forestry production, significant threats to low-lying populations in equatorial areas, and the risk of infectious disease and heat-stress related deaths. By the middle of the twenty-first century, average annual river runoff and water availability should increase by 10–40% in high latitudes and in typically wet tropical regions, but water will decrease by 10–30% over currently drier areas. Thus, drought-stricken areas will likely increase in spatial extent. Conversely, heavy precipitation events will increase in frequency, causing higher risk of flash floods in areas already vulnerable to such incidents. Water availability will be severely impacted in those regions dependent on freshwater that is sourced by snow cover and glaciers, since both of these freshwater resources will become severely limited during the century.

Many ecosystems are expected to collapse due to the combination of warming, changes in water availability, wildfire, insects, ocean acidification, land use changes, pollution, and over exploitation of natural resources. The ecosystems most likely to be affected first are those where the rate of change will be fastest or most severe—such as in the Arctic (when the ice cap begins to disappear each summer) and in ocean fisheries near already vulnerable coral reefs. Over the course of the twenty-first century, net carbon uptake by terrestrial ecosystems is likely to climb for a time, but then weaken or even reverse, accelerating the rate of warming due to positive feedback. Major shifts in the ranges of many species, and major decreases in the biodiversity of most global ecosystems would be expected if global warming exceeds about 4°F. As noted in Chapter 4, this will take place during the 2030s on the current warming trajectory in the Northern Hemisphere, and somewhat later in the Southern. The progressive acidification of the oceans due to the absorption of carbon dioxide is expected to negatively impact all marine shell forming organisms and the large numbers of species that depend upon them as either a food source or a source of cover.

Crop productivity is projected to increase slightly due to climatic factors at mid to high latitudes until mid-century, when the excess heat will begin to harm yield. At lower latitudes, which are dominated by developing countries of lower adaptive ability, crop yields are probably already being negatively impacted by climate factors, and this trend will worsen as the warming proceeds. Crops in all world areas are expected to be negatively impacted by changes in rainfall patterns, not only in terms of drought, but also heavy precipitation events, and the possible increased frequency of severe storms. Forestry is generally expected to be somewhat positively impacted by the expected changes, though the amount of area devoted to managed forestry may need to shrink to accommodate food production. Aquaculture and fisheries will be adversely affected due to the combination of warming, acidification, and other stressors (such as hypoxia, see the discussion of that topic later in this chapter).

Coastal systems will face increasing risks due to erosion, sea-level rise, possible increased frequency and severity of tropical cyclones, and higher heat, all of which will be made more serious by the generally low adaptive capacity of these regions. Widespread mortality in coral reefs will impact those coastal communities that rely upon the fisheries linked to those coral systems. Salt marshes and mangrove swamps will be harmed by sea-level rise, and many millions more people are expected to experience flood every year due to sea-level rise. The heavily populated "megadeltas" of Asia and Africa will be among the hardest hit parts of the planet, as will be communities on small islands with especially vulnerable topography.

The net economic impacts on society will be negative across the planet, but some regions are expected to fare better than others—generally those not experiencing drought or other water-related stresses, such as flooding. Upland areas in the mid- to high-latitude regions would appear to be the "best place to be," but these are exactly the areas that are least populated today. For instance, in the United States, population growth is fastest in coastal regions, urban areas near rivers, and "sunbelt" stops like Las Vegas, which are heavily dependent on external sources of water.

The health-related impacts of climate change in the twenty-first century will be widespread around the planet but will inevitably strike the developing nations of the world the hardest, where adaptive capacity is generally the lowest. Malnutrition is a likely outcome from difficulties in food production, and there will be increased deaths due to heat waves, floods, storms, fires, and drought. Diarrheas will become more widespread and there is an expectation of increased frequency of cardiorespiratory diseases due to ozone. Finally, there will be altered spatial distribution of some infectious disease vectors. The present-day fear of an imminent avian flu pandemic is probably just the first of many such incidents as the planet continues to warm and grows increasingly crowded.

Another type of impact discussed by IPCC is the relative likelihood of changes in various categories of extreme weather events, such as heat waves, heavy precipitation events, drought, tropical cyclones, and extremely high sea levels. They go on to discuss the possibility of "catastrophic" climate events, such as a sudden shutdown in the MOC ocean current in the North Atlantic, but IPCC concludes that a large abrupt transition is very unlikely during the twenty-first century.

The final portions of the IPCC impacts summary are devoted to a discussion of society's current adaptive capacity and what factors are limiting adaptation. The IPCC concludes that there are barriers to adaptation primarily associated with cost, informational, attitudinal, and behavioral issues. In all, the IPCC seems to dwell more on the catastrophic impacts of coming climate change, rather than on the more practical issues of what humanity might be doing to try to prepare for it.

Impacts by Region

As in the reports on climate change, the IPCC documents contain detailed discussions of the most likely impacts to each region of the planet. I've made an attempt to just hit the high points in the brief regional summaries listed below.

Africa

As soon as 2020, between 75 and 250 million Africans are expected to encounter some form of water stress, and agricultural production—already challenging to many African countries—will become even more difficult. Some countries are expected to see yield decreases of 50% or more. Africa is expected to be the hardest hit of all continents in terms of the human cost of global warming.

Asia

Freshwater availability in much of central, southern, eastern, and Southeast Asia will decrease as glacier melt continues to decline and eventually disappear for many watersheds. More than one billion people will be impacted by this loss of water resource. Food production will fare somewhat better than in Africa, but yields could be negatively impacted by the middle of the century. Populations in the mega-deltas will suffer from the already mentioned effects of climate change on these regions.

Australia and New Zealand

Drought is expected to continue as an increasingly common concern in southern and eastern Australia, threatening most of agriculture. New Zealand, which is at higher latitude and not as drought prone, may actually have improved growing conditions. Fisheries in these areas, such as those based on the Great Barrier Reef, are expected to suffer considerably due to the impacts of warming and ocean acidification. In all, the region is expected to be fairly resilient to climate change, largely due to relatively smaller human populations and substantial adaptive resource. This may help to explain Australia's initially cool response to the Kyoto Protocol.

Europe

As a whole, Europe will see a continued loss of snow cover and glacial-fed stream-flows, but there will be a significant gradient in the impacts of climate change running from north to south—with the northern Europeans generally benefiting and the rest of the continent faring progressively worse as one proceeds south and east. The greatest threats in the south will be reduced water availability, wildfires, and lower crop productivity. In central and eastern Europe, the threats will be summertime drought and heat waves. Northern Europe may benefit from longer growing seasons. Compared with less developed parts of the world, conditions in Europe will be quite tolerable, at least for moderate levels of global warming.

Latin America

The greatest change in Latin America will be in the eastern Amazon basin, which will shift from a tropical rainforest to a savannah, with concomitant changes in vegetation. Biodiversity will be negatively impacted by this change. Increased desertification and salinization are expected in areas already vulnerable to these threats. As in other world areas, freshwater resources will suffer from the decrease of snow cover and glacial melt water. In general, however, Latin America is expected to do much better than their neighbors across the Atlantic.

North America

Agriculture in the United States and Canada is expected to benefit, especially in the more northern portions of the growing regions and especially where rainfall is already sufficient to avoid irrigation. The western portions of the continent that already suffer deficits in annual precipitation are expected to see this trend increase in severity, which is a challenge because of the rapid rate of population growth in many of these areas. Forest fire risk will continue on its recent trend of increased severity. Certain coastal communities are vulnerable to the threats posed by increased tropical cyclone activity and rising sea level. New Orleans and nearby communities would still top that list.

Subsequent to the release of the IPCC assessment during 2007, the US

Climate Change Science Program issued its own forecast in May 2008. This report was produced by a large panel of American experts under contract to the Commerce Department. These authors largely agree with the sentiments of the international group. To the extent that they do differ with IPCC forecasts, it would be on the side of concluding that changes appear to be happening faster than recognized by the United Nations experts—as I have argued in this book.

Polar Regions

Human populations in the Arctic were the first to be confronted by the changes brought on by man-made global warming. For the most part, the changes seem manageable and could be easily viewed as an improvement over the previous, "natural" climate. The increased navigability has provided certain economic opportunities for these communities. The current ecosystems, however, are likely to be severely disrupted as major species (such as polar bears and other mammals) are displaced.

Small Islands

The human communities on many small islands will be severely threatened by the impacts of climate change. In addition to the obvious impact of sea-level rise, ocean acidification has the potential to completely eliminate the current source of livelihood for many of these people, with the very real possibility that some will be forced to relocate entirely. Similarly, there is an increased risk of freshwater shortages due to changes in rainfall patterns and increased rates of evaporation. These factors could put additional pressure on such populations to move.

Graphical Summary of Impacts

The IPCC summary document also provides a very useful graphic for visualizing the overall impacts of varying degrees of warming on the systems that are most likely to be affected. I have added English temperature units (degrees Fahrenheit) to the original graphic (Table SPM–1 in the IPCC report) and included it here as Figure 5.1. It shows the specific effects on water, ecosystems, food, coasts, and health to particular global mean temperature increases over the range of 0–10°F. To the bottom of the graphic,

I have added the years that these temperatures would be reached according to the current Northern Hemisphere land surface warming trend (as shown in Figure 4.3).

Ocean Acidification

As explained in Chapter 2, proper understanding and modeling of the oceans is absolutely essential for accurate climate modeling. Most of the excess energy that is currently being trapped by greenhouse gases actually ends up in the ocean. The global network of oceanic currents is also a dominant factor in determining many weather patterns around the planet. But there is also a vast quantity of really important chemistry and biology going on in the oceans as well. Oceanic phytoplanktons contribute roughly half of the biosphere's net primary photosynthetic production of organic material, making them a key link in the global carbon cycle. Each day, more than a hundred million tons of carbon is fixed in the upper ocean. There is now direct evidence of lower carbon fixation in the ocean at higher sea surface temperatures (Behrenfeld et al., 2006). This is a positive feedback process that is not in most of the current IPCC models.

But of even greater concern to the health of the ocean is the newly discovered phenomenon of ocean acidification. This is undoubtedly the greatest threat the oceans now face from the continued atmospheric release of excess carbon dioxide by humanity. This newly identified issue recently caused the United Kingdom's Royal Society, which has a storied record as one of the world's premier scientific organizations, to get actively involved. In 2005, they published a comprehensive and exceedingly well written (as only the Brits seem able to do) discussion of the current state of scientific knowledge about ocean acidification.

The Society's analysis first summarizes the key known facts. About half of all the carbon dioxide that man has released through the combustion of fossil fuels has been deposited in the oceans. This has led to a reduction of the pH of surface seawater of 0.1 units, equivalent to a 30% increase in the concentration of hydrogen ions. At current trends, the pH would be expected to fall by an additional 0.5 units during the twenty-first century, implying a trebling in the concentration of hydrogen ions, an acidity not seen in the oceans for possibly 300 million years. Perhaps most critically, this rate of change is probably one hundred times faster than any known geochemical change in the entire history of the oceans.

The most alarming change in chemistry brought on by this increased acidity will be the dissolution of calcium carbonate, the material used by virtually all marine organisms for their shells. The values given in the previous paragraph represent global averages and there is extensive geographic variation, meaning that certain regions will experience impacts much more rapidly. The Society also points out that ocean acidification is essentially irreversible over meaningful time scales for response, because of the very slow mixing and sedimentation processes that are responsible for providing the natural buffering capacity of the seas. The timescales for these natural processes are on the order of tens of thousands of years.

The Society therefore concludes that drastic reductions in current levels of carbon dioxide emission appear to be the only practical way to minimize the risk of large-scale and long-term changes to the oceans. They further point out that, though the science around the biological impacts of these changes is only in its infancy, the early indications suggest that many marine organisms are at immediate risk, especially corals, mollusks, and many microscopic organisms that rely on the process of calcification to produce their shells.

They recommend a major new international effort to study these effects and to determine the extent to which this biological damage represents a major—yet currently ignored—positive feedback effect in the overall global warming process. This would happen if the seawater chemistry changes become severe enough to measurably decrease the ability of the world's oceans to carry out the photosynthetic processes that help to fix carbon dioxide. Currently, such processes represent a major sink for the anthropogenic carbon dioxide (the extra amount of this material being produced by man's combustion of fossil fuels).

The Society concludes that the increased fragility and sensitivity of marine ecosystems needs to be taken into consideration during the development of any policies that relate to their conservation, sustainable use and exploitation, or the communities that depend on them. As noted earlier, the discussion ends with one last plea for world leaders to set new, aggressively lower targets for carbon dioxide emissions, and that all possible approaches should be considered to prevent carbon dioxide from reaching the atmosphere, "no option that can make a significant contribution should be dismissed."

The concerns of the Royal Society are echoed in a number of other recent publications by marine scientists, which point out that the oceans are already under tremendous stress due to pollution, overfishing, and hypoxia, the last of which I will discuss next. Many of these same scientists are now speaking out, stating there is an urgent need to develop plankton monitoring programs worldwide in order to act as sentinels for detecting the changes anticipated with continuing changes in temperature, carbon dioxide levels, pH, and dissolution of calcium carbonate.

Hypoxia

The last environmental challenge that I will touch upon in this chapter, and only quite briefly, is hypoxia. Hypoxia (as the Greek suggests) is a condition of oxygen depletion. But among environmental scientists it generally refers to a seasonal phenomenon now being observed where large rivers enter oceans. The leading explanation for hypoxia is that excessive nutrients (mainly nitrogen and phosphorous) from agricultural activities and municipal water treatment facilities combine with warm weather to produce intense algal blooms that consume dissolved oxygen and then die, sink to the bottom, and decompose, in the process creating a "dead zone" of nearly 10,000 square miles.

In 2001, the US Government formed the Mississippi River/Gulf of Mexico Watershed Nutrient Task Force, which issued a document entitled, "An Action Plan for Reducing, Mitigating, and Controlling Hypoxia in the northern Gulf of Mexico." In early 2008, the same task force issued the "Gulf Hypoxia 2008 Action Plan." This document gives an excellent summary of what is currently known scientifically about seasonal hypoxia in the northern Gulf of Mexico. The hypoxic zone is now occurring each summer, just offshore of the point where the Mississippi River discharges all runoff from the 41% of the contiguous United States that it drains.

In 2007, the measured size of this northern Gulf hypoxic zone was 20,500 square kilometers (7,900 square miles), which is about the size of the state of Massachusetts. This is far larger than the stated goal of the Task

Force, which is to bring the five-year moving average down to less than 5,000 square kilometers (about 1,900 square miles)—more like the size of Rhode Island. Global warming is expected to make hypoxia all the more challenging: Hypoxia tends to increase the amount of stratification between the incoming fresh surface water and the colder sea water beneath, because ever warmer surface water will be that much less dense and slower to sink—–as if we didn't have enough bad things to blame on global warming!

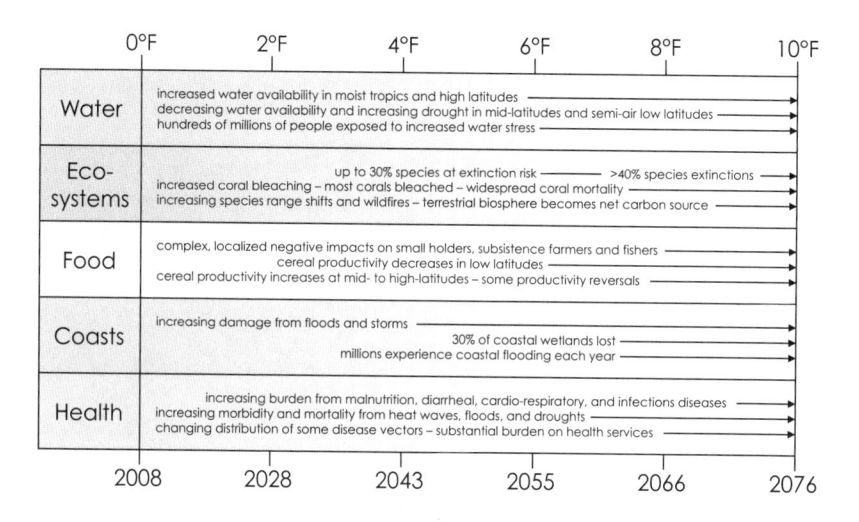

	0°F 2°F 4°F 6°F 8°F 10°F
Water	increased water availability in moist tropics and high latitudes decreasing water availability and increasing drought in mid-latitudes and semi-arid low latitudes hundreds of millions of people exposed to increased water stress
Eco-systems	up to 30% species at extinction risk —— >40% species extinctions increased coral bleaching – most corals bleached – widespread coral mortality increasing species range shifts and wildfires – terrestrial biosphere becomes net carbon source
Food	complex, localized negative impacts on small holders, subsistence farmers and fishers cereal productivity decreases in low latitudes cereal productivity increases at mid- to high-latitudes – some productivity reversals
Coasts	increasing damage from floods and storms 30% of coastal wetlands lost millions experience coastal flooding each year
Health	increasing burden from malnutrition, diarrheal, cardio-respiratory, and infections diseases increasing morbidity and mortality from heat waves, floods, and droughts changing distribution of some disease vectors – substantial burden on health services
	2008 2028 2043 2055 2066 2076

FIGURE 5.1 Illustrative examples of global impacts projected for climate changes associated with different amounts of global warming during the twenty-first century (based on Table SPM–1 from the IPCC WGII Fourth Assessment Report Summary for Policymakers).

Chapter Six

"WHAT SORT OF PEOPLE OUGHT YOU TO BE?"

As I ANSWER THIS QUESTION, I will not spend any time advocating for any particular political position or commenting on the various public policies that are now being debated, such as carbon cap and trade systems or higher fossil fuel taxes. I agree with the recently released assessment of NASA's Jim Hansen (2008), who concludes that humanity would need to reduce atmospheric carbon dioxide down to a level of 350 PPM in order to avoid "catastrophic effects." Because we are still seeing accelerating carbon dioxide levels of about 385 PPM and climbing, and there is no global consensus on any plan to reduce emissions, I believe this to be the most rational position. Mitigation is unlikely, and the task before humanity is to simply adapt as best we can to what will become increasingly inhospitable conditions. For followers of Christ, such a world will present unprecedented opportunities to care for our fellow man—especially those in Africa and other parts of the planet forecasted to suffer the most. As you will see, I am calling on fellow believers to abandon lives of mindless energy consumption—not because it will measurably slow global warming—but because it will help preserve our testimony before a perishing world. I find the clear implication of all the scientific facts is that these are the Last Days, meaning this is our final opportunity to present Jesus as the only means of salvation for every soul upon this planet. It's time we focused on that task!

I will answer the "so what?" question from my perspective, that of a practicing environmental scientist who happens to have a biblical worldview.

When Peter asked the partly rhetorical question of this chapter title, his stated context was very similar to where we apparently now find ourselves, "the elements will be destroyed with intense heat, and the earth and its works will be burned up. Since all these things are to be destroyed in this way, what sort of people ought you to be in holy conduct and godliness, looking for and hastening the coming of the day of God" (2 Pet. 3:10–12). I will try to bring Peter's admonition up to date, while at the same time attempting to reconcile it with the other concerns that should fill our hearts as environmentally aware, born-again followers of Jesus: (1) the tension between God's sovereignty and man's free will, (2) the scientific limits to what can actually be accomplished to slow the current warming trajectory, (3) our God-given responsibility to serve as stewards of Earth, (4) God's commandment to love our neighbors as ourselves, and (5) the Great Commission (to proclaim the gospel to the very ends of the Earth). This is a very tall order, but I'll give it a whirl!

Is Our Free Will Sufficient to Reverse Global Warming?

As should be clear to the reader by now, I am convinced that all prophecy in the Bible will be ultimately fulfilled, including those prophetic passages that I believe have parallels in the theory of man-made global warming. One might then ask, "If God has already predetermined that the Earth is going to burn up, why should we bother trying to avoid global warming?" I find this question to be analogous to one posed oratorically by Paul in the book of Romans, "Why does He still find fault? For who resists His will?" (Rom. 9:19). My first answer is the same as Paul's, "who are you, O man, who answers back to God?" (Rom. 9:20).

But this seems a bit intellectually shallow, and it ducks the real dilemma. For those who have studied philosophy extensively, perhaps the mystery of reconciling God's sovereignty and man's free will seem less like a challenge and more like an insoluble Gordian knot in a tired old rope, but I find it fascinating. The mere existence of fulfilled prophecy in the biblical and historical record provides direct evidence of at least some God-imposed limitations on our free will. Nevertheless, God certainly considers the measure of free will that we have been given to be sufficiently "free" in order for Him to righteously condemn those of us who refuse His offer of salvation.

The eternal torment of the second death in the lake of fire hardly seems fair if the unrepentant individual could have been spared through just one sovereign act of grace. However, the mere fact that we are capable of expressing such lofty thoughts seems ample evidence that the infinite God who created us will always do what is right. As for the unresolved tension in this intellectual puzzle, I rest in the comfort of knowing there is much in God's vast universe that I may not ever fully comprehend, at least not while housed in this mortal coil.

Meanwhile, the scientist in me is still curious about the following question.

Is it Scientifically Possible for Us to Stop Global Warming?

This is a loaded question. I believe the theoretical answer is yes, but the practical answer is no, because it would require either many millennia or an immediate, massive infusion of capital into unproven fuel and energy systems—and we don't have enough time, money, resolve, or new technology. Having answered in this way, I know that I am completely at odds with the majority of United Nations bureaucrats, who say that we can and we must stop global warming, and that the United States should pay for it! This is also the answer given by the trophy-laden Al Gore, with an Oscar in one hand and a Nobel Peace Prize in the other. However, I disagree. In saying so, I know it's not going to win me a trip to Hollywood or Oslo, but I'm planning on a more pleasant final destination!

As I showed in Chapter 4, the actual warming trajectory that the Earth is now following is at approximately twice the warming rate predicted by the IPCC models. Those same models say that global warming would continue for another several decades even if we had miraculously stopped adding any new greenhouse gases to the atmosphere back in the year 2000. The IPCC models also predict continued warming for many centuries in all of the realistic greenhouse gas emission scenarios they have presented. I'm not sure how the IPCC spin doctors will tweak their messaging as their scientists begin to awaken to the fact that their models are not keeping pace with the actual rate of global warming. But if I had to guess, I'd predict they will argue there is an even stronger need for immediate United Nations intervention—despite the fact that some of their more honest scientists may acknowledge the practical futility of trying to reverse the warming. We'll see.

So let's ask a follow-up question.

SHOULD WE TAKE REASONABLE STEPS THAT MIGHT SLOW GLOBAL WARMING?

Many leaders have decided that the answer to this question is "Yes!" They have grown weary of hearing scientists and politicians from either end of the spectrum shout at each other about the underlying theory, so they have decided to just do something…anything! For most, this has meant formulating altruistic mission statements and establishing long-term targets (typically out to the year 2050), that call for massive reductions in greenhouse gas emissions but don't specify how to do it. This is exactly what Governors Schwarzenegger (California) and Crist (Florida) have done here in the United States.

Even though most politicians (and perhaps most people) would answer yes to the above question, that doesn't mean they would necessarily agree on what constitutes "reasonable steps." Would they all move closer to work into a tiny studio apartment, trade in their minivan for a used *Prius,* and observe a weekly "Carbon Sabbath," as I have done? I doubt it. Would they be willing to replace incandescent light bulbs as they burn out in their homes with fluorescent energy savers and spend an extra five grand for a hybrid engine in their new SUV the next time they go car shopping? Perhaps they would. But will these types of voluntary measures reverse or even meaningfully retard global warming? They probably will not. I believe such actions, however "warm" they may make our hearts feel, will be ultimately ineffective due to the quickening pace of emission growth in China, India, and other world areas—not to mention the huge fuel and energy infrastructure in the developed world that just keeps on emitting. The Energizer Bunny comes to mind!

So why do it?

THE HIGH COST OF INACTION

In the case of man-made global warming, I believe that doing nothing (the "business as usual" scenario) is no longer a viable political or personal option. I say this from both a practical "worldly" perspective and from a biblical perspective. Like it or not, and as I write this, I can think of several personal friends who still vehemently disagree, a solid majority of the world's intellectual leaders have come to accept the theory of man-made global warming as true. So to argue for inaction puts one in the precarious position of trying to defend an "unlimited right to pollute." This includes not only the unlimited consumption of fuel and energy—but that of all the commercial goods that have a large "hidden" carbon footprint such as fresh food imported by air. In the eyes of the world, attempting to defend such a "right to wanton consumption" is increasingly being seen as arrogant and selfish.

From a biblical perspective, I would argue that refusing to repent of excessive energy consumption is sinful on many counts. It is a form of idolatry (a violation of the second commandment), as it places the right of unlimited personal consumption upon the highest pedestal within our hearts, which is the rightful place of God. It is prideful, because it reflects an attitude that values personal freedom above all else. Third, it displays an attitude of disdain—rather than love—toward the fellow inhabitants of this planet. The vast majority of our global neighbors do not live in societies with the extremely high adaptive capacity and other resources that we have been blessed with here in the United States. Man-made global warming will harm our neighbors far before and far more than it ever will harm us. In taking this position, I recognize that some will view my admonition as nothing more than self-righteous criticism. To those whom I have thus offended, I humbly acknowledge the sin that remains within my own life and pray that you will accept my call as coming from a broken jar of clay who prays daily for the infilling of the Holy Spirit.

We serve a God who calls us to love Him with all of our heart, mind, and strength—and our neighbors as ourselves. Paul wrote mournfully of the many people he knew who had failed to do this, "they are enemies of the cross of Christ, whose end is destruction, whose god is their appetite, and whose glory is in their shame, who set their minds on earthly things" (Phil. 3:18–19). Continued mindless consumption of unlimited energy in a time such as this is simply not a loving act. John's words seem quite timely for this age, "Do not love the world, nor the things in the world. If anyone loves the world, the love of the Father is not in him. For all that is in the world, the lust of the flesh and the lust of the eyes and the boastful pride of life, is not from the Father, but is from the world. And the world is passing away, and also its lusts; but the one who does the will of God abides forever" (1 John 2:15–17).

REAL LOVE DOES SOMETHING!

As so eloquently stated by Paul in his first letter to the believers at Corinth, the ability to express Godly love (αγαπη "agape" love) toward the LORD and all of our global neighbors—including not just friends but all enemies—is the supreme act of obedience to which we are all called, "If I speak with the tongues of men and of angels, but do not have love, I have become a noisy gong or a clanging cymbal. And if I have the gift of prophecy, and know all mysteries and all knowledge; and if I have faith, so as to remove mountains, but do not have love, I am nothing. And if I give all my possessions to feed the poor, and if I deliver my body to be burned, but do not have love, it profits me nothing" (1 Cor. 13:1–3). All other spiritual gifts are temporal, but love is not. For we will all have the opportunity to continue expressing that love for the Lord and for one another for the eternity that shall continue after this world has passed away. In the meantime, we are called to express real love, and to do it now, while we are still here.

Love such as this does not sit idly by while untold billions live out their lives of quiet desperation on the broad path that ultimately leads to personal destruction. Real love calls for action. As explained by Jude, in the brief note that became the penultimate book of our Bible, "May mercy and peace and love be multiplied to you. Beloved, while I was making every effort to write you about our common salvation, I felt it necessary to write to you appealing that you contend earnestly for the faith which was once for all delivered to the saints" (Jude 2–3). Contending "earnestly" may mean setting aside some of the worldly comforts that come with excessive energy consumption and spending a bit more personal effort on God's chief intended purpose for our time here.

What is our purpose?

"Becoming an R12 Christian"

One of my favorite Christian broadcasters, Chip Ingram, has recently coined a phrase that I believe is very relevant to the present discussion. It's called, "becoming an R12 Christian," where "R12" stands for Romans 12, which begins with the following exhortation, "I beseech you therefore, brethren, by the mercies of God, that ye present your bodies a living sacrifice, holy, acceptable unto God, which is your reasonable service. And be not conformed to this world, but be ye transformed by the renewing of your mind, that ye may prove what is that good, and acceptable, and perfect, will of God" (Rom. 12:1–2, KJV). In Chip's use of this terminology, he has asked his listeners to make an affirmative commitment to hand over complete control of their lives to God and to His mission upon this earth, which is to gather a people unto Himself, "Come to Me, all who are weary and heavy-laden" (Matt. 11:28).

I believe Ingram's challenge is relevant to the present question because much of our consumption (at least here in the United States) is completely discretionary and usually selfish. It is direct evidence that we are allowing ourselves to be conformed to the world, by following its pattern of mindless consumption, rather than being transformed by the renewing of our minds through acts of direct worship of God and sacrificial service to others.

Just one book later in the Bible, Paul comments on the nature of the true liberty that we have in Jesus, "All things are lawful for me, but not all things are profitable" (1 Cor. 6:12). In this context, Paul was speaking primarily about dietary and sexual indulgences, but it is only a small extrapolation to extend his logic to the issue of energy consumption. Having been reconciled to God through Jesus' work of atonement on the Cross, we are no longer under the threat of punishment under the Law, so we are free to do "all things," but at what cost? If we, as followers of Jesus, are seen hanging out at a brothel, drinking ourselves silly at a bar, or conspicuously burning thousands of gallons of jet fuel in a private jet—then we will find ourselves hurting the cause of Jesus, "by sinning against the brethren and wounding their conscience" (1 Cor. 8:12).

The Bottom Line—The Finish Line

So this all leads to my biblical answer to the "Why do it?" question. I choose to refrain from excessive energy consumption, because I believe it brings a more disciplined and focused life. It is a life that seeks to honor God and provide a better witnessing opportunity to a world full of lost souls, who happen to be living on a planet cursed by our careless behavior. Later in 1 Corinthians, Paul puts it this way, "I do all things for the sake of the gospel" (1 Cor. 9:23), and continues with this, "I buffet my body and make it my slave, lest possibly, after I have preached to others, I myself should be disqualified" (1 Cor. 9:27), and "though I am free from all men, I have made myself a slave to all, that I might win the more" (1 Cor. 9:19).

For Paul, life was a race, "forgetting what lies behind and reaching forward to what lies ahead, I press on toward the goal for the prize of the upward call of God in Christ Jesus" (Phil. 3:13–14). Paul's was not a race to consume, as it seems to be here in the United States, but a race to win souls for Jesus. And Paul most assuredly won that race. Near the end of his time on earth, sitting in a dank and cold Roman prison, he was able to say, "I have fought the good fight, I have finished the course, I have kept the faith; in the future there is laid up for me the crown of righteousness, which the Lord, the righteous Judge, will award to me on that day; and not only to me, but also to all who have loved His appearing" (2 Tim. 4:7–8).

Walk the Walk—"Woe to You Hypocrites!"

Another reason for responding to the issue of global warming by leading a life of reduced energy consumption is that it addresses one of the greatest criticisms that unbelievers have for those of us who profess to be followers of Jesus. They think we are hypocrites. Unfortunately, they are often right. Although it isn't directly related to the cause of Christ, Al Gore's hypocrisy on the issue of his personal energy consumption (private jets, the size of his electric bill, etc.) has been the fitting butt of many a joke. May it never be said of us, and may each instead remember to "walk in a manner worthy of the calling with which you have been called" (Eph. 4:1).

It's worth remembering that it is not only unbelievers who chastise hypocrites. Jesus reserved His harshest criticism for Israel's religious leaders—

whom He saw as hypocrites—showering them with a tirade of seven woes, the last of which contains particularly venomous language, "Woe to you, scribes and Pharisees, hypocrites!... You serpents, you brood of vipers, how shall you escape the sentence of hell?" (Matt. 23:29–33). He also implied Hell will be the ultimate destination for all such hypocrites when He said of the evil servant, "cut him in pieces and assign him a place with the hypocrites; weeping shall be there and the gnashing of teeth" (Matt. 24:51). James was somewhat more reserved in his language while addressing fellow believers, but his letter strongly emphasizes the importance of leading a life of actions that are consistent with our fine Christian doctrine, "But prove yourselves doers of the word, and not merely hearers who delude themselves" (James 1:22).

Paul pointed out that hypocrisy is entirely inconsistent with the true spirit of love and service that should characterize the heart of a true follower of Jesus, "Let love be without hypocrisy. Abhor what is evil; cling to what is good. Be devoted to one another in brotherly love; give preference to one another in honor; not lagging behind in diligence, fervent in Spirit, serving the Lord; rejoicing in hope, persevering in tribulation, devoted to prayer, contributing to the needs of the saints, practicing hospitality" (Rom. 12:9–13). In subsequent letters, Paul also wrote, "Therefore be imitators of God, as beloved children; and walk in love, just as Christ also loved you, and gave Himself up for us, an offering and a sacrifice to God as a fragrant aroma" (Eph. 5:1–2), and this, "walk by the Spirit, and you will not carry out the desires of the flesh" (Gal. 5:16). And John also reminded us that "the one who says he abides in Him ought himself to walk in the same manner as He walked" (1 John 2:6).

Let Your Light Shine

As Christians, we are called to let our light shine. And that doesn't mean every light in the unoccupied rooms of our homes! Jesus said, "You are the light of the world. A city set on a hill cannot be hidden. Nor do men light a lamp and put it under a peck-measure, but on the lampstand; and it gives light to all who are in the house. Let your light shine before men in such a way that they may see your good works, and glorify your Father who is in heaven" (Matt. 5:14–16). In our times, the good works may mean setting aside some of our modern conveniences for the sake of maintaining the purity of our testimony, as we follow Jesus and lead others to Him who said, "I am the light of the world; he who follows Me shall not walk in the darkness, but shall have the light of life" (John 8:12).

This is a high calling, but it is the one that we are asked to answer. At times, it may mean sacrifice—*real* sacrifice—as Jesus also said:

> "If anyone wishes to come after Me, let him deny himself, and take up his cross, and follow Me. For whoever wishes to save his life shall lose it; but whoever loses his life for My sake and the gospel's shall save it. For what does it profit a man to gain the whole world, and forfeit his soul? For what shall a man give in exchange for his soul? For whoever is ashamed of Me and My words in this adulterous and sinful generation, the Son of Man will also be ashamed of him when He comes in the glory of His Father with the holy angels" (Mark 8:34–38).

Stay Humble!

One of the real dangers here is that we would mistakenly pursue asceticism for the sake of asceticism, and thereby fall into the prideful trap of becoming self-righteous in our criticism of those around us—including our Christian brothers and sisters—who choose not to follow our example of self denial. It is this self-righteous form of judgment that Jesus taught against when He said, "Do not judge lest you be judged" (Matt. 7:1). As has been repeatedly emphasized here, the point of leading a less consumptive life is to preserve our testimony, to humbly acknowledge our own sinful contribution to the world's problems, and to direct our neighbors to the only real way of salvation from the Earth's plight, Jesus.

The person whose heart is governed by such biblical humility will not be so prideful to take the seat of judgment, which is reserved for God alone. The Bible is consistent throughout its pages on extolling the virtues of this Godly humility. In the short book of Micah, there is a familiar passage that speaks of this, "And what does the LORD require of you but to do justice, to love kindness, and to walk humbly with your God?" (Mic. 6:8). Jesus explained it this way, "whoever exalts himself shall be humbled; and whoever humbles himself shall be exalted" (Matt. 23:12).

Peter also praises humility, and the blessings that will follow from remaining steadfast in faith, despite the challenges that are sure to come from the triune enemies of our souls (the world, our sinful flesh, and Satan):

> "Humble yourselves, therefore, under the mighty hand of God, that He may exalt you at the proper time, casting all your anxiety upon Him, because He cares for you. Be of sober spirit, be on the alert. Your adversary, the devil, prowls about like a roaring lion, seeking someone to devour. But resist him, firm in your faith, knowing that the same experiences of suffering are being accomplished by your brethren who are in the world. And after you have suffered a little while, the God of all grace, who called you to His eternal glory in Christ, will Himself perfect, confirm, strengthen and establish you" (1 Pet. 5:6–10).

Encourage One Another!

The other thing to remember in these days of increasing peril is that there is strength in numbers. As born-again followers of Jesus, we are outnumbered in this world and this is likely to remain true, as Jesus warned, "Enter by the narrow gate; for the gate is wide, and the way is broad that leads to destruction, and many are those who enter by it. For the gate is small, and the way is narrow that leads to life, and few are those who find it" (Matt. 7:13–14). Nevertheless, a "few" out of the nearly seven billion people now on this planet is a big number, "a great multitude, which no one could count" (Rev. 7:9).

I know from personal experience that there is much to be gained by gathering with other believers, and the benefits are likely to increase, as explained by the writer of Hebrews in two passages, "encourage one another day after day, as long as it is still called, 'Today,' lest any one of you be hardened by the deceitfulness of sin" (Heb. 3:13), and this, "Let us hold fast the confession of our hope without wavering, for He who promised is faithful; and let us consider how to stimulate one another to love and good deeds, not forsaking our own assembling together, as is the habit of some, but encouraging one another, and all the more, as you see the day drawing near" (Heb. 10:23–25).

As that day draws ever nearer, there is at least one threat lurking close by.

THE CURSE OF PROSPERITY

It could rightly be said that the greatest danger facing the Church in the United States today is the "curse of prosperity." As we grow increasingly comfortable with the lush lifestyle powered by our extraordinary consumption of fuel and energy—magnificent megachurches, cozy first and second homes, plush belongings, electronic toys, ever fancier vehicles, and globe-trotting vacations—there is a very real risk of us falling into the idolatry of worshipping the Creation rather than the Creator.

It would be wrong to say that God does not want to bless us by meeting our material needs, "Give us this day our daily bread" (Matt. 6:11). But He is thoroughly against greed, and the Bible consistently teaches the virtues of charity, "When you reap the harvest of your land, moreover, you shall not reap to the very corners of your field, nor gather the gleaning of your harvest; you are to leave them for the needy and the alien" (Lev. 23:22). This particular law is repeated two other times, once earlier in Leviticus (19:9), and in Deuteronomy (24:19). This law figured prominently in the wonderful story of Ruth, the lovely heroine of the book bearing her name, who gleaned as a widow and an alien in the fields of Boaz near Bethlehem and who subsequently becomes the great grandmother of none other than King David.

David grew up in those same hills surrounding Bethlehem, and, through the anointing work of the Holy Spirit, he found himself in Jerusalem during his life-changing middle-age crisis, when the prosperous and sumptuous life he was leading in the palace helped cause him to fall into the great sin of his life—the adulterous affair with Bathsheba and the murder of her husband, Uriah the Hittite. It was the curse of prosperity that helped cause David to fall. It should be a warning to us all. Jesus spoke of this danger in the parable of the sower, "And the one on whom seed was sown among thorns, this is the man who hears the word, and the worry of the world, and the deceitfulness of riches choke the word, and it becomes unfruitful" (Matt. 13:22).

But Jesus did not stop there. It ought to send chills down the spine of every follower of today's so-called "prosperity gospel," when they read Jesus' following warning, "Truly I say to you, it is hard for a rich man to enter the kingdom of heaven. And again I say to you, it is easier for a camel to go through the eye of a needle, than for a rich man to enter the kingdom of God" (Matt. 19:23–24). Luke's gospel records Jesus adding the following warning, "But woe to you who are rich, for you are receiving your comfort in full. Woe to you who are well-fed now, for you shall be hungry" (Luke 6:24–25). Luke also gave us Jesus' story of a "certain rich man" and Lazarus (Luke 16:19–31) that speaks of the eternal dangers inherent in selfishly pursuing earthly comfort while not having mercy on the "poor man laid at the gate."

James also spoke of this futility in chasing after these temporary pleasures, "But let the brother of humble circumstances glory in his high position; and let the rich man glory in his humiliation, because like flowering grass he will pass away. For the sun rises with a scorching wind, and withers the grass; and its flower falls off, and the beauty of its appearance is destroyed; so too the rich man in the midst of his pursuits will fade away" (James 1:9–11). Later in his letter, James was even more graphic with his imagery:

> "Come now, you rich, weep and howl for your miseries which are coming upon you. Your riches have rotted and your garments have become moth-eaten. Your gold and your silver have rusted; and their rust will be a witness against you and will consume your flesh like fire. It is in the last days that you have stored up your treasure! Behold, the pay of the laborers who mowed your field, and which has been withheld by you, cries out against you; and the outcry of those who did the harvesting has reached the ears of the Lord of Sabaoth. You have lived luxuriously on the earth and led a life of wanton pleasure; you have fattened your hearts in a day of slaughter" (James 5:1–5).

Peter echoed the same sentiment when he wrote, "Beloved, I urge you as aliens and strangers to abstain from fleshly lusts, which wage war against the soul. Keep your behavior excellent among the Gentiles, so that in the thing in which they slander you as evildoers, they may on account of your good deeds, as they observe them, glorify God in the day of visitation" (1 Pet. 2:11–12).

In the last of the messages to the angels of the seven churches (in Laodicea, which many interpret as a message to the modern Church of our day), Jesus issued a fiery rebuke:

> "I will spit you out of My mouth. Because you say, 'I am rich, and have become wealthy, and have need of nothing,' and you do not know that you are wretched and miserable and poor and blind and naked, I advise you to buy from Me gold refined by fire, that you may become rich, and white garments that you may clothe yourself, and that the shame of your nakedness may not be revealed; and eyesalve to anoint your eyes, that you may see. Those whom I love, I reprove and discipline; be zealous therefore, and repent" (Rev. 3:16–19).

Later in the same book, John wrote of a "great harlot" who never did repent (a professing, but false church?), and he wrote of her horrible fate, "To the degree that she glorified herself and lived sensuously, to the same degree give her torment and mourning; for she says in her heart, 'I sit as a queen and I am not a widow, and will never see any mourning'" (Rev. 18:7).

But she will. Let us not be counted with her.

Teach Us to Number Our Days

Of the 150 Psalms given to us in the Bible, only one is ascribed to Moses, Psalm 90, in which it is written, "Who understands the power of Thine anger, and Thy fury, according to the fear that is due Thee? So teach us to number our days, that we may present to Thee a heart of wisdom" (Ps. 90:11–12). The linking of wisdom to being careful to make the best use of our time is a theme repeated by Paul, "Therefore be careful how you walk, not as unwise men, but as wise, making the most of your time, because the days are evil" (Eph. 5:15–16).

Solomon, renowned for his wisdom, spoke of the organized manner in which God has ordered the time sequences in all of Creation, "There is an appointed time for everything. And there is a time for every event under heaven" (Eccles. 3:1). And I would argue it is now time for those who profess Jesus to boldly stand up for the biblical truth about the coming destruction of this planet with man-made global warming playing a key role in the process. To those who are ashamed to step forward and simply state what is plainly written in the Bible, I would counsel them in the same way that Mordecai spoke to his uncle's daughter, Queen Esther, some 2,500 years ago, "For if you remain silent at this time, relief and deliverance will arise for the Jews from another place and you and your father's house will perish. And who knows whether you have not attained royalty for such a time as this?" (Esther 4:14).

Be Alert!

As we consider how we ought to live our lives in these days of coming global calamities, I believe it is also biblical to maintain an attitude of constant anticipation, as instructed by Jesus, "Be on your guard, that your hearts may not be weighted down with dissipation and drunkenness and the worries of life, and that day come on you suddenly like a trap; for it will come upon all those who dwell on the face of the earth. But keep on the alert at all times, praying in order that you may have strength to escape all these things that are about to take place, and to stand before the Son of Man" (Luke 21:34–36). Mark's gospel quotes Jesus on a different occasion, "Therefore, be on the alert – for you do not know when the master of the house is coming, whether in the evening, at midnight, at cockcrowing, or in the morning – lest he come suddenly and find you asleep. And what I say to you I say to all, 'Be on the alert!'" (Mark 13:35–37). Paul expressed a very similar thought, "And this do, knowing the time, that it is already the hour for you to awaken from sleep; for now salvation is nearer to us than when we believed. The night is almost gone, and the day is at hand. Let us therefore lay aside the deeds of darkness and put on the armor of light" (Rom. 13:11–12).

But this alert attitude should not be accompanied by panic or fear, for "we know that God causes all things to work together for good to those who love God, to those who are called according to His purpose" (Rom. 8:28). And a little later in this same book, Paul wrote of the certainty that we shall endure whatever hardship shall come, "For I am convinced that neither death, nor life, nor angels, nor principalities, nor things present, nor things to come, nor powers, nor height, nor depth, nor any other created thing, shall be able to separate us from the love of God, which is in Christ Jesus our Lord" (Rom. 8:38–39).

Last Words on Conduct

Paul's quick summary of Christian conduct, which appears at the end of his first letter to the church at Thessalonica, seems a great way to begin summarizing all that has been said in this chapter:

> "Live in peace with one another. And we urge you, brethren, admonish the unruly, encourage the fainthearted, help the weak, be patient with all men. See that no one repays another with evil for evil, but always seek after that which is good for one another and for all men. Rejoice always; pray without ceasing; in everything give thanks; for this is God's will for you in Christ Jesus. Do not quench the Spirit; do not despise prophetic utterances. But examine everything carefully; hold fast to that which is good; abstain from every form of evil" (1 Thess. 5:13–22).

Finally, from the pen of Paul, we read this, "And let us not lose heart in doing good, for in due time we shall reap if we do not grow weary. So then, while we have opportunity, let us do good to all men" (Gal. 6:9–10). My prayer is that each of you would find encouragement in these words, as the time of our final deliverance from this dying planet draws near.

Chapter Seven

Recap and Benediction

Within these few pages, I have barely scratched the surface of what both modern science and the Bible have to tell us about the reality of man-made global warming. If you want to know more about the future and what God's desire is for your personal life, then I would direct you to one of the four Bibles in the average American home, often serving merely as a bit of forgotten décor. For those wanting to learn more about what modern science has to say, I have included an annotated bibliography in Appendix 1, immediately following this chapter. This chapter is intended to serve as a brief recap of what has been said, and I conclude with what I hope is taken as a parting word of genuine encouragement.

My hope is that this book will have added a new voice to the cacophonous chorus that now reverberates around the world concerning the reality of man-made global warming and what we, as individuals, are called to do in response to the growing threat that many of us believe it represents. As I described in Chapter 1, "Modern Science vs. the Bible," the perspective that I have offered here was one that gave both modern science and an ancient text (specifically, the Bible) approximately equal time and opportunity to comment upon this issue.

As a practicing environmental scientist and a born-again follower of Jesus, this seems entirely appropriate to me, given the profound and universal significance of this issue for all of us who find ourselves sharing this planet. For many of you, I know this perspective may have seemed strange and almost impermissible in today's otherwise open marketplace of ideas. To those of you who have retained an open mind and stayed on the journey contained within these pages, I thank you for your openness and willingness to wade patiently though my attempts to tackle this immense topic. But hopefully I have been able to effectively convey the simple reality that I have come to believe—that the theory of man-made global warming is true, and that widespread disruptions of this planet's ecosystems are inevitable because of it. I find this scientific fact to be a profound defense of the Bible, whose authors have consistently taught this same truth for thousands of years, whereas modern science has only recently "discovered" it.

In Chapter 2, "Climate Science," I described the process by which modern science has methodically come to this conclusion. It has done so by the same scientific method formulated centuries ago by the Christian, Roger Bacon—through an iterative process of empirical observation, deductive reasoning, model and hypothesis development, experimental design, objective analysis of observations, and modification of the model and/or hypotheses as appropriate. In the specific case of modern climate science, this process began at the very end of the nineteenth century, when the first hypotheses appeared in the scientific literature about the potential greenhouse effect of combustion-derived gases. The twentieth century development of this area of science was then characterized by both quantum leaps forward and missteps sideways and backwards as the scientific peer review process erratically played itself out, as it generally does. Along the way, climate science earned (or to be more fair, was rather saddled by) a reputation for bold and sometimes outrageous predictions. Chief among these were the ice age scares of the early 1970s. This notoriety remains a screeching monkey on the back of any person, whether a scientist or not, who possesses the combination of temerity and courage (or is it foolishness!) to step forward and offer a new analysis to the already immense and ever-growing body of literature on this subject. I am keenly aware of that primate's presence as I humbly add these words to that mountain of material.

In Chapter 3 "Climate Change in the Bible," I jumped out of the frying pan of the climate science debate and into the fiery world of biblical interpretation. As a scientist by academic training and member of the laity with respect to theological matters, I have endeavored to restrict my biblical discussions to the "main and plain" things of the Bible when it comes to climate. I am indebted to two unnamed fellow believers in Christ who graciously agreed to review early drafts of this manuscript and provide their comments, which I have sought to fully incorporate. I thank them both for their careful review, but I take full responsibility for any and all errors that may have been made here—not only in the analyses of Scripture, but in all other portions of this book. For me, the obvious "bottom line" of the Bible with respect to the theory of man-made global warming is that the theory is not only entirely consistent with the ancient Scripture, it is actually foretold in surprising detail.

I find several passages in the Book of Revelation to be particularly explicit in this regard. Revelation 11 states that "the time has come…to destroy those who destroy the earth" (Rev. 11:18). The clear implication is that man's actions are directly responsible for the coming catastrophes. Revelation chapters 8 and 16 contain a series of global "natural" disasters that mirror the latest computer model predictions of global climate changes and impacts with an eerie and exquisite synchronicity. If this doesn't cause the current unbeliever in the authority of biblical texts to reexamine whether they might actually be divinely inspired by an omniscient and omnipotent God, then I would have to question that person's intellectual integrity and willingness to accept truth for what it is…true!

In Chapter 4, "Predicted Climate of the Twenty-First Century," I switched back to modern science to describe the key climate predictions for the remainder of this century, as described in the most recent set of IPCC reports. At the end of the chapter, I gave my own analysis of the available temperature data, which suggests that the current climate models are actually too sluggish, with observed temperatures accelerating at a rate that is about twice as fast as the IPCC models. I showed that the current rate of temperature increase across the land surfaces of the Northern Hemisphere is 0.9°F per decade and is closely following an accelerating curve that would add a sweltering 16°F degrees to mean temperatures by the end of the twenty-first century.

What will these warmer temperatures mean to humanity and the Earth, as we now know it? In Chapter 5, "Predicted Climate Change Impacts," I again relied mainly on the IPCC reports to describe the very troubling events now predicted to take place later in this century as a result of climate change, including widespread species extinctions, human disease, and famine. Among the most troubling of all the phenomena predicted to take place is the process of "ocean acidification." According to modern marine scientists, this change in oceanic chemistry will lead to the dissolution of calcium carbonate, the key component forming the shell of numerous marine organisms. It seems highly unlikely that "evolution" could occur at a rate sufficient to salvage these organisms or otherwise compensate for this drastic alteration of marine chemistry.

The sad and unavoidable conclusion is that oceanic food webs, already crippled by overfishing and the direct toxic effects of man-made pollutants, will become increasingly stressed and will eventually collapse, thereby explicitly fulfilling the words penned by the Apostle John on the Isle of Patmos over 1,900 years ago, when he wrote of the ocean, "and it became blood like that of a dead man; and every living thing in the sea died" (Rev. 16:3). As I explained in Chapter 3, the blood of a dead man is characterized by elevated CO_2 and dropping pH—exactly what is happening to the oceans now, with accelerating rapidity.

Having painted this admittedly bleak picture, which is the same whether you take it from the current thinking of modern scientists or from the plainly stated words of the Bible, I gave a series of biblical answers to the "so what?" question posed in the title of Chapter 6, "What Sort of People Ought You to Be?" This is the same rhetorical question posed by Peter when he affirmed the same doomed fate of the earth that had been foretold by all of the great Jewish prophets before him. Many of the answers I gave to this "So what?" question come down to nothing more than simple, common-sense measures of trying to downsize our personal lives, which does not have a hefty price tag—indeed, quite the opposite. Most answers result in a significant savings, making it possible for those of us who name the name of Jesus to be in a better financial position to contribute to the Great Commission.

Such answers may not sound too different from the ones we hear from the many secular environmentalists among us. But the basis for the answers I have offered is generally very different from that of most such advocates, and it is the same as that used throughout the book. I stand on ground leveled by the latest thinking of modern science but always tempered by what I know to be the truth of Scripture—I call this a scientifically-informed biblical worldview.

When Peter answered the "so what?" question he wrote that we ought to exhibit, "holy conduct and godliness, looking for and hastening the day of God" (2 Pet. 3:11). When Peter said "hastening the day," it is clear that he could not have possibly meant that we should do everything we can to hasten the destruction of the planet—quite the contrary. He was personally called by his recently risen Savior, Jesus, to proclaim the saving message of the gospel to the very ends of the earth. He had been personally instructed by Jesus that this global distribution of the gospel message must happen first, and then the end would come. This is what Peter meant when he urged his readers to "make haste."

It is my personal belief that we now live in that "evil and adulterous" generation to which Jesus referred when He discussed the Last Days. Despite widespread skepticism and mocking in the West, and even the beginnings of persecution of the Church, the gospel is now reaching the uttermost corners of the world, as Jesus said must happen before the end comes. There is an unprecedented rate of growth in the body of believers in China, Africa, Latin America, and even nations currently dominated by Islam, where persecution is characterized by outright genocide, as in the Darfur region of Sudan.

But I believe it is perfectly biblical to obey this mission of spreading the gospel while living a personal lifestyle of reduced consumerism and consumption. I don't believe we should sacrifice the gospel on the altar of environmentalism, for that ultimately equates with worshipping the Creation rather than the Creator. God alone is worthy of our worship. However, if called, the follower of Jesus should be willing to do what the rich young ruler was apparently not willing to do, "go and sell your possessions and give it to the poor, and you shall have treasure in heaven; and come, follow Me" (Matt. 19:21).

I pray that we would all show compassion toward our fellow man in these coming troubled times by being willing to reduce our mindless consumption of excess fuel and energy and instead invest our maximum possible effort in

Jesus' call to proclaim His message of salvation to the ends of the earth while there is still time. For He has promised: "Yes, I am coming quickly" (Rev. 22:20).

May the Lord shield you from what's to come—Amen.

Appendix One

Annotated Bibliography

ANONYMOUS (2003). Living landscapes: Chapter 3: The changing atmosphere. Royal BC Museum, http://www.livinglandscapes.bc.ca/thompok/env-changes/atmos/ch3.html. This site gives an excellent overview of the various manmade atmospheric pollutants, the relative magnitude of their natural and atmospheric sources, and their impacts. It also presents a simple explanation of the processes responsible for the production of photochemical smog, acid rain, and dry deposition of acidic pollutants.

Anonymous (2007). Chapter 6. Geochemical Cycles. http://wwwas.harvard.edu/people/faculty/djj/book/bookchap6.html. This is a portion of a basic textbook that explains the fundamental chemical relationships involved in the mass balance of carbon dioxide in the atmosphere, carbonate chemistry in the oceans, carbon uptake by the terrestrial biosphere (land plants), and the overall preindustrial carbon cycle.

ANONYMOUS (2007). Cooling off the global warming debate. Special mailing of Center for Reclaiming America for Christ, an outreach of Coral Ridge Ministries, April 2007.

ANONYMOUS (2007). Evangelicals and global warming. Special mailing of Center for Reclaiming America for Christ, an outreach of Coral Ridge Ministries, April 2007. A short sidebar article describes the controversy stirred in late 2006 and early 2007 when Rick Warren (of *Purpose-Driven Life fame*) and eighty-five other evangelical leaders issued a call to action for Christians to curb greenhouse gas emissions via a print and TV ad campaign in *The New York Times,* on Fox News, and elsewhere.

Anonymous (2007). Light at the end of the tunnel. *Nature,* 445:567. This is a commentary on the latest IPCC report, and it contains the following bold statement, "The IPCC report has served a useful purpose in removing the last ground from under the skeptics' feet, leaving them looking marooned and ridiculous." For the remaining "true" skeptics in the scientific community, there are plenty of other "fighting words" here, and the piece is unapologetic in lamenting the success that the opponents of the "scientific consensus" have been winning in the political realm, especially in the United States. But the article is fairly evenhanded in the way that it exposes the naked hypocrisy of European governments, which blast out calls for mandatory emissions by the evil Americans while at the same time planning airport expansions that will enable a further tripling in air travel– –the fastest-growing source of emissions, and one not capped by the Kyoto Protocol. Presumably, this exception is intended to make sure that the thousands of IPCC scientists and policy experts don't feel quite so ridiculous and self-conscious as they fly around the world from one global warming conference to another.

ANONYMOUS (2007). Still waters: the global fish crisis. *National Geographic,* April 2007, pp 33–69. The article describes, in the magazines typically alarmist tone, how overfishing has decimated multiple fish populations to the point where countries dependent on seafood appear to be doomed to rely on hot dogs and hamburgers in the near future.

ANONYMOUS (2007). The big thaw. *National Geographic,* June 2007, pp 56–71. In a tone similar to that of the previous entry, this article sounds a siren over the unexpected and frightening rate at which glaciers and polar ice continue to melt.

ANONYMOUS (2007). Acid rain makes some Shenandoah National Park streams unfavorable to fish. US Geological Survey, News Release, 24 September 2007. This report from USGS highlights the vulnerability of the streams of this National Park to the negative effects of acid rain on fish communities. The main contributors to acid rain are the sulfur and nitrogen containing pollutants during the combustion of certain fossil fuels. In addition to highlighting the current plight of these streams, the press release goes on to forecast that the future habitability of these streams will be quite poor, with a greater than 90% probability of at least one acid episode every year for four consecutive years.

ANONYMOUS (2008). *Gulf Hypoxia Action Plan 2008,* Mississippi River/Gulf of Mexico Watershed Nutrient Task Force.

AUTIO CF (2005). What is your worldview? *Answers in Genesis,* http://www.answersingenesis.org/docs 2005/0502worldviewasp. The author is a retired Major General from the US Air Force, who argues the most simplistic definition for a biblical worldview is to "have the mind of Christ" (Phil. 2:5), which he takes to mean that one would "think like Christ; love like Christ; act like Christ; walk like Christ: have the humility, patience, longsuffering and all of the other Galatians 5:22–26 fruit of the Spirit." According to Autio, such a worldview dominated American thought until about the year 1900, when America became more and more influenced by the "modernist" thinking that had been bubbling up out of Europe during most of the nineteenth century. He quotes recent (2003) Barna Group polling data that suggest only 4% of all American adults and only 9% of professing born-again Christians have such a perspective. In the same poll, only half of America's Protestant pastors (51%) were reported to have such a biblical worldview. Of course, the results of such polling data are very sensitive to the definitions used, but the bottom line is that American church as a whole seems strangely silent on the hard line that Jesus, Paul, and all the apostles took on issues such as divorce, pornography, sexual deviancy, and the inerrancy and infallibility of Scripture.

BEGLEY S (2007). The truth about denial. *Newsweek*, 13 August 2007, pp 21–29. An unapologetic and unbalanced tirade about the "Grand Conspiracy" that she alleges is being led by ExxonMobil and other unnamed industry collaborators against the theory of man-made global warming. Although some of the more outrageous claims against ExxonMobil were subsequently blunted by the Samuelson retraction a week later, there is still plenty of ammo here, most of it aimed against the current Bush administration, although even Bush "the elder" takes some of the shrapnel. Of course, Clinton's failure to defend the Kyoto treaty is immediately defended as the only politically expedient course of action. All in all, it is a typical mainstream media piece on the theory of man-made global warming. It is very high on hype and attacks Republicans and corporations, but there is not very much truth to go around.

BEHRENFELD MJ, O'Malley RT, Siegel DA, McClain CR, Sarmiento JL, Feldman GC, Milligan AJ, Falkowski PJ, Letelier RM, Boss ES (2006). Climate-driven trends in contemporary ocean productivity. *Nature*, **444**:752–755. The authors state that oceanic phytoplankton contribute roughly half of the biosphere's net primary production, making it a key link in the global carbon cycle. Each day, more than a hundred million tons of carbon is fixed in the upper ocean. The authors present direct evidence of lower carbon fixation at higher sea surface temperature. They therefore conclude that future temperature increases will inevitably alter net air-sea exchange of carbon dioxide, fishery yields, and dominant basin-scale biological regimes.

BEHRENFELD MJ, Worthington K, Sherrell RM, Chavez FP, Strutton P, McPhaden M, Shea DM (2006). Controls on tropical Pacific Ocean productivity revealed through nutrient stress diagnostics. *Nature*, **442**:1025–1028. The authors report the results of new methodologies for directly measuring the impact of various stressors on the productivity of bloom-forming diatoms. The data reported in this particular paper are only weakly related in any direct manner to the issue of global climate change, but they do represent a significant advance in the science of measuring impacts on the biological productivity of the world's oceans at "unprecedented" spatial scale. Since the article was only approved in July 2006, it betrays the infancy of the monitoring techniques that have even been developed for detecting biological changes in the world's oceans.

BERAKI AF (2005). Master's thesis, University of Pretoria, South Africa. This is an interesting example of the type of regional climate modeling assessments that are now being used in many world areas to guide public policy decisions. In this particular case, Beraki investigates the high likelihood of intense drought in Eritrea, a region already heavily impacted by the expanding Sahara and the stress this brings to the people groups of that war-torn region.

BIRGER J (2007). The great corn gold rush. *Fortune Magazine*, 29 March 2007. Birger reports on the huge spike in corn grain prices and planting intentions for the 2007 growing season in the American Midwest. Ethanol production is expected to consume 26% of the 2007 corn crop, and that number is expected to reach 36% by 2008.

BOURNE JK (2007). Green dreams: making fuel from crops could be good for the planet—after a breakthrough or two. *National Geographic*, October 2007, pp 38–59. This article is packed with information about the biofuels revolution now sweeping the planet. Some of the crops discussed as new fuel sources include corn-based ethanol, sugarcane-based ethanol, canola- and soy-based biodiesel, cellulosic ethanol from prairie grasses, and algae. The article highlights a common view that some of these biofuels, as currently produced, may have only very limited benefits with respect to their ability to reverse current global warming trends.

Caldeira K, Wickett ME (2003). Anthropogenic carbon and ocean pH. *Nature,* **425**:365. This brief article reports that the expected drop in ocean pH (0.7 units) will be more extreme than any level known during the past 300 million years. They do not delve into the biological impact of such a change but only comment in passing that impacts on coral reefs, calcareous plankton, and other organisms whose skeletons or shells contain calcium carbonate may be particularly affected.

CARRIER F (2007). Governments must deal with greenhouse gases: US Supreme Court, *Agence France Presse,* 2 April 2007. Typical of many media reports, and especially those from France, the lead sentence of the news articles presents this court decision as a "blow to US Present George W. Bush." The case itself was brought against EPA by a coalition led by Massachusetts, several other states, cities, and environmental organization. The specific legal question was whether greenhouse gases fit within the Clean Air Act's broad overall definition of an "air pollutant." The court found that it did. The Government unsuccessfully argued that even if they imposed regulation of such gases, there would be no direct benefit due to the infinitesimal nature of the theoretical effect of such emission controls, either on motor vehicles or electricity-generation plants.

CHRISTENSEN JH, Hewitson B, Busuioc A, Chen A, Gao X, Held I, Jones R, Kolli RK, Kwon W-T, Laprise R, Magna Rueda V, Mearns L, Menendez CG, Raisanen J, Rinke A, Sarr A, Whetton P (2007). Regional climate projections. In: *Climate Change 2007: The Physical Science Basis. Contribution of Working Group I to the Fourth Assessment Report of the Intergovernmental Panel on Climate Change* [Solomon S, Qin D, Manning M, Chen Z, Marquis M, Averyt KB, Tignor M, and Miller ED (eds.)]. Cambridge University Press: Cambridge, UK and New York, NY, pp 847–940.

CRICHTON M (2004). *State of Fear.* New York, NY: Avon Books, 672 pp. Crichton presents a spellbinding story of global ecoterrorism and the spy-like heroes trying to prevent them, with a clear goal of exposing what Crichton believes to be the grand conspiracy that is behind the theory of man-made global warming. To the end of the novel he added a pair of interesting appendices (see next entry) and an impressive annotated bibliography, which summarizes the extensive body of environmental science texts that he consulted for three years while writing the novel.

CRICHTON M (2004). Author's message. In *State of Fear.* New York, NY: Avon Books. At the conclusion of his novel, Crichton takes several pages to summarize what he believes to be the "take-home" messages from his three years of researching the global warming issue. He concludes that we really can't know how much, if any, of the current warming is man-made, and launches into various attacks against those who exploit this uncertainty to their own advantage, financial or otherwise. Of course, Crichton himself is guilty of this, which he sarcastically acknowledges in the last of several "bullet" points, "Everybody has an agenda. Except me."

DIFFENBAUGH NS, Pal JS, Trapp RJ, Giorigi F (2005). Fine-scale processes regulate the response of extreme events to global climate change. *PNAS,* **102**:15774–8.

ENKVIST P-A, Naucler T, Rosander J (2007). A cost curve for greenhouse gas reduction. *The McKinsey Quarterly,* 2007 Number 1. This detailed economic investigation claims to be the first of its kind in its stated scope of including all relevant greenhouse gases, all sectors, and all geographic regions. In addition to this global breadth, it spans three time horizons (2010, 2020, 2030), three target greenhouse gas concentrations (400, 450, 500 PPM CO2 equivalent), and presents the cost of each known abatement measure (cost units are euros per ton of carbon). For the global economy, the analysis concludes that achieving the 450 PPM target would cost about 500 billion euros in 2030, but uncertainties in the analysis suggest the cost could be as high as 1,100 billion euros annually, which they estimate to be about 1.4% of global GDP at that time.

FARRELL AE, Sperling D (2007). *A Low-Carbon Fuel Standard for California: Part 1: Technical Analysis,* University of California, Berkeley and University of California, Davis, 29 May 2007. This detailed technical report is an outgrowth of the broad Global Warming Solutions initiative now under way in the State of California. The Low Carbon Fuel Standard (affectionately known as the LCFS) calls for a reduction of at least 10% in the *carbon intensity* of California's transportation fuels by the year 2020. The term, carbon intensity, is deceptively simple to explain and virtually impossible to measure with precision. It refers to the totality of the global warming impacts associated with the production, transport, storage, and use of the fuel—including indirect effects, such as the land use changes that accompany many forms of biofuel production. The units of measure are grams of carbon dioxide equivalents per unit of fuel energy delivered to the vehicle, adjusted for inherent differences in the in-use energy efficiency of different fuels. Because it is defined in this way, it is not enough to merely document the chemical identity of the fuel—one must document the new global warming impacts of all of the activities that went into producing it. As such, it is a far-reaching and ambitious regulation that is certain to encounter legal challenges, as did a recent set of new mileage efficiency standards set by the state.

FARRELL AE, Sperling D (2007). *A Low-Carbon Fuel Standard for California: Part 2: Policy Analysis,* University of California, Berkeley and University of California, Davis, 1 August 2007. This policy analysis builds upon the technical case presented a month earlier and makes a series of specific recommendations concerning the LCFS: (1) the LCFS should apply to all gasoline and diesel used in California for transportation, including freight and off-road applications, and should allow participation by providers of non-liquid fuels (electricity, natural gas, propane, and hydrogen); (2) differences in the drive train efficiency of diesel and gasoline engines should be accounted for and heavy and light duty diesel fuels should be treated differently; (3) recommendations are made for the baseline and targets for reductions in carbon intensity; (4) the LCFS should be imposed upon providers and importers of fuel; (5) greenhouse gas emissions associated with the production of fuels should be included; (6) a default and opt in system should be employed, with "pessimistic" default values and the burden of proof on the provider to justify lower life cycle global warming characteristics for the alternative fuel; (7) providers should be permitted to trade or bank (hold) LCFS credits, but no borrowing should be allowed; (8) providers should be allowed to comply by paying a fee, which is different from paying a fine for noncompliance; (9) third-party auditors should be used to verify that claimed credits are accurate; (10) empirical data should be used to establish drivetrain efficiency adjustment factors; (11) offsets from within the transportation sector should be permitted but not from outside the transportation sector; (12) carbon capture and storage technologies that are safe, adequately monitored, and directly related to the supply of transportation energy should be included in the LCFS; (13) Life Cycle Analysis is judged to be adequate for the purpose of establishing an initial quantitative framework for the LCFS, but a program to improve such methods should be implemented; (14) develop a non-zero estimate of the global warming impact of direct and indirect land use change for crop-based biofuels for use in the first several years of the LCFS and participate in international efforts to improve this methodology; (15) initially keep LCFS separate from the California auto efficiency standards (AB1493) but consider integration at a later date; and (16) coordinate the LCFS with AB32 (the larger Global Warming Solutions policy), which may include a "hard cap" on emissions. The analysis concludes with a series of "softer" recommendations such as paying attention to environmental justice and sustainability issues, additional credits for "more innovative" low carbon fuels, interim program review (2013), and of course, a call for more research!

FEELY RA, Sabine CL, Fabry VJ (2006). *Carbon Dioxide and Our Ocean Legacy,* Pew Charitable Trusts, April 2006, 4 pp. This is a brief "glossy" intended to help prompt public (and funding?) interests in the subject of carbon dioxide impacts on the world's oceans. The picture on page 1 looks like an outtake from Disney's "Finding Nemo," except that anemone is green, rather than pink. A somewhat outdated pie chart (based on 2002 figures) still shows the United States as the largest greenhouse gas emitter. The inner pages introduce the chemistry relationships of dissolved carbon dioxide, pH, and calcium carbonate dissolution in highly simplified form. It concludes with several scary sections entitled, "Reefs at Risk," "Cracks in the Food Chain," and "Fish – It's What's For Dinner." The final sentence is in boldface, "The message is clear: excessive carbon dioxide poses a threat to the health of our oceans."

GLEICK J (1988). *Chaos: Making a New Science.* New York, NY: Penguin Books, 352 pp. Gleick's book came out at a unique moment of the computer age when it was just becoming possible to produce color prints of the fanciful, fractal-based images that are now relatively commonplace and form the underlying theory for the amazingly realistic computer-generated video animations that are used to either enhance or completely produce modern motion pictures. As he tells the history of the science that led to these discoveries, Gleick begins by spending a good deal of time explaining the so-called "Butterfly Effect," which still leads many to mistakenly believe (as I once did) that the prediction of long-term climatic responses to thermal forcing (as by excess atmospheric carbon dioxide) is not possible.

GORDON C (2007). Tracking glacial activity in Norway with photogrammetry software. *Imaging Notes,* Spring 2007, pp 24–29. Gordon describes a research study in which remote sensing data were used to fine-tune estimates of how quickly glaciers will continue to melt under various climate change scenarios. Such data will contribute to improved predictions of the rate of sea level rise during the twenty-first century.

GORE A (2006). *An In Convenient Truth: The Planetary Emergency of Global Warming and What We Can Do About It.* Emmaus, PA: Rodale, 328 pp.

HANSEN J, Sato M, Kharecha P, Beerling D, Masson-Delmotte V, Pagani M, Raymo M, Royer DL, Zachos JC (2008). Target atmospheric CO2: where should humanity aim? preprint posted at http://www.realclimate.org, accessed on 6 May 2008. The lead author on this posting has become a lightning rod for drawing the fiery bolts of the skeptic community, possibly because he continues to point out the overly conservative and sluggish nature of IPCC computer modeling projections. Hansen has dual affiliations with Columbia University and NASA's Goddard Institute for Space Studies in New York, NY. This particular paper, which has not yet drawn much media attention here in the United States, calls for a drastically lower target for atmospheric carbon dioxide: 350 PPM. Achieving such a level would require Draconian changes in human behavior, such as phasing out all global use of coal (without carbon capture and storage), revolutionary new forestry and agricultural practices for sequestering carbon, and complete halt on further exploitation of new oil reserves. Needless to say, none of the current global proposals for reducing GHG emissions come even close to this target. Without such a reduction, however, Hansen concludes that the planet is doomed to "catastrophic effects."

HAYS GC, Richards AJ, Robinson C (2005). Climate change and marine plankton. *Trends in Ecology and Evolution,* **20**:337–344. The authors present a review of the interactions between climate change and plankton communities, focusing on systematic changes in plankton community structure, abundance, distribution, and phenology over recent decades. They also examine the potential socioeconomic impacts of these plankton changes, such as the effects of bottom-up forcing on commercially exploited fish stocks. They discuss the crucial roles that plankton might have in dictating the future pace of climate change via feedback mechanisms. They conclude that there is an urgent need to develop plankton monitoring programs worldwide in order to act as sentinels for detecting the changes anticipated with continuing changes in temperature, carbon dioxide levels, pH, and dissolution of calcium carbonate.

HOPKIN M (2007). Climate skeptics switch focus to economics. *Nature,* **445**:582–83. Hopkin discusses how mainstream skeptics of climate change have recently altered their tactics. They no longer focus on questions surrounding the accuracy of the global temperature measurements. Instead, they are now questioning the eventual economic impacts of government strategies such as the current push toward corn-ethanol based biofuels.

IGLESIAS-RODRIGUEZ MD, Brown CW, Doney SC, Kleypas J, Kolber D, Kolber Z, Hayes PK, Falkowski PG (2002). Representing key phytoplankton functional groups in ocean carbon cycle models: coccolithophorids. *Global Biogeochemical Cycles,* 16(4), 1,100, doi:10.1029/2001GB001454. The authors present a very complicated, multifaceted assessment of the potential for climate-induced changes in the blooms of coccolithophorids, an important class of primary calcifying phytoplanktons. The discussion eventually muses about the possibility that future trends in ocean chemistry may begin to favor one class of organisms but is frustratingly short on specific testable predictions for all of its length.

IPCC (2001). *IPCC Special Report on Emissions Scenarios,* Intergovernmental Panel on Climate Change, http://www.grida.no/climate/ipcc/emission/093.htm.

IPCC (2007). *Climate Change 2007: The Physical Science Basis, Summary for Policymakers.* Geneva, Switzerland: IPCC, 21 pp.

IPCC (2007). *Climate Change 2007: Climate Change Impacts, Adaptation and Vulnerability, Summary for Policymakers.* Geneva, Switzerland: IPCC, 23 pp.

ISAACSON W (2007). *Einstein: His Life and Universe.* New York, NY: Simon & Schuster, 680 pp. Isaacson presents the man, his scientific theories, and his understanding of God, all in a comprehensive way that has only become possible with the recent release of Einstein's personal letters. Perhaps most astonishing is the fact that Einstein's entire career was that of an "outsider," partly due to his Jewish heritage in a Europe that was to be overrun by the Nazis but also due to an innate bent against all forms of authority and conformity extending even to his stubborn refusal to ever accept the prevailing quantum mechanics descriptions of the subatomic world that his own scientific work had helped to create.

JOHNSON J (2007). EPA boosts ethanol: air pollution is allowed to increase for new and expanding ethanol biorefineries. *Chemical & Engineering News,* 30 April 2007, p 27. The article describes a relaxation of air quality standards for the new ethanol biorefineries that are springing up all across the nation's heartland, based primarily on the use of corn as the feedstock to produce ethanol. The article states that capacity will be increasing over the next few years to twelve billion gal/yr over the current level of 5.8 billions gal/yr.

JOHNSON J (2007). *Building green. Chemical & Engineering News,* 16 July 2007, pp 11–15. The article highlights the increasing trend toward "green" construction of both homes and businesses in the United States. Since some 40% of the greenhouse gas emissions in the United States are directly attributable to the energy expended to light, energize, heat, and cool such structures, this represents an obvious opportunity for reduction efforts.

KLEYPAS JA, Feely RA, Fabry VJ, Langdon C, Sabine CL, Robbins LL (2006). *Impacts of Ocean Acidification on Coral Reefs and Other Marine Calcifiers: A Guide for Future Research,* report of a workshop held 18–20 April 2005, St. Petersburg, FL, sponsored by NSF, NOAA, and the US Geological Survey, 88 pp. The report is highly detailed and well written, containing the combined efforts of nearly fifty scientists who met for an initial April 2005 workshop and subsequently produced an impressive summary of the disturbing impacts that increasing concentrations of oceanic carbon dioxide are likely to have on the marine calcium carbonate system, coral reefs, and other marine calcifiers. It describes the current state of scientific knowledge about these systems and what changes are anticipated over the twenty-first century as carbon dioxide inputs to the oceans continue to increase and pH continues to drop, leading to the dissolution of calcium carbonate, the basis for much of marine life, such as coral reefs. Major impacts are possible, but the authors' main conclusion appears to be a plea for more research funding.

LeGrande AN, Schmidt GA, Shindell DT, Field DV, Miller RL, Koch DM, Faluvefi G, Hoffman G (2006). Consistent simulations of multiple proxy responses to an abrupt climate change event. PNAS, 103:837–842 (2006).

Marchese J (2007). Al Gore is a greenhouse gasbag. *Philadelphia Magazine,* February 2007. Marchese uses an interview with Robert Giegenback, a provocative geology professor at the University of Pennsylvania, in an attempt to convince readers that Al Gore's *Inconvenient Truth* is nothing more than a scare tactic. Geigenback's key argument is that sea-level rise, which he argues is the best way to gauge global warming, proves there are no immediate threats. He goes on to accuse Gore of blatantly attempting to position himself as a viable candidate for the 2008 election. He also points out that if carbon dioxide is really the problem that some scientists say it is, then it is far too late to stop it because of the exponentially increasing emissions from China and India. However, his personal attitude seems largely dominated by the lengthy timescales of his chosen field of science, "The Earth was fine before we got here, and it'll be fine long after we're gone."

McKibben B (2007). Carbon's new math: to deal with global warming, the first step is to do the numbers. *National Geographic,* October 2007, pp 32–37. This essay presents global warming as the "greatest test humans have yet faced." It lists a series of options that would reduce annual carbon emissions by a billion metric tons each. The strategies fall in the following categories: conservation or efficiency, carbon capture and storage, low-carbon fuels, and renewables and biostorage.

Melcer R (2007). WU to build center for $55 million. *St. Louis Post Dispatch,* 5 June 2007, p. C1. Melcer reports on a new collaboration between Washington University (St. Louis), the University of Missouri (Columbia), and the Danforth Plant Science Center to develop the International Center for Advance Renewable Energy and Sustainability, or I-CARES. The goal is discovery of innovations that can reduce carbon dioxide emissions and slow or halt global warming. One of the key focus areas will be to find new ways of making fuel from existing sources, such as ethanol or coal, while also developing new ones.

MILLS E (2006). "The Role of NAIC in Responding to Climate Change," testimony to the National Association of Insurance Commissioners winter meeting, San Antonio, TX, 8 December 2006. Mills is a staff scientist from the Lawrence Berkeley National Laboratory, with considerable expertise in the area of natural disasters and the cost of responding to them. In this invited testimony, Mills gives a detailed assessment of the insurance cost implications of the many predictions coming from the latest IPCC reports. He highlights several types of current insurance coverage where the availability of insurance is likely to suffer in coming years: crop losses, soil subsidence, permafrost melt, flood and mudslide damage, drought-induced crop losses, wildfire, infectious disease, coastal erosion, and property losses due to inland severe storms (ice, tornadoes, etc.). He also highlights the increasing number of wildfires in the western United States, the continuing high number of powerful tropical cyclone damage, and the dramatic death toll of the great European heat wave of 2003.

MORELLE R (2007). Neanderthal climate link debated. *BBC News*, 13 September 2007. This news story highlights an ongoing debate among anthropologists and paleoclimatologists concerning potential links between the demise of Neanderthals (*Homo neanderthalensis*) and a sudden cooling of northern Europe, approximately 24,000 years ago. The debate centers on some Neanderthal remains found in three caves along the coast of southern Iberia, and whether these fossil remains prove or disprove a climate link.

NCDC (2007). *Global surface temperature anomalies.* National Climatic Data Center, National Environmental Satellite, Data, and Information Service, National Oceanic and Atmospheric Administration, US Department of Commerce, http://www.ncdc.noaa.gov/oa/climate/research/anomalies/anomalies.html. This is a link to downloadable files containing the latest monthly and annual global temperature anomalies. The monthly data are available separately for land and sea, and for the Northern and Southern Hemispheres.

ODLING-SMEE L (2007). What price a cooler future? *Nature,* **445**:582–83. Odling-Smee writes of a raging debate among global economists surrounding a key October 2006 report by Nicholas Stern, a senior British civil servant and former vice president of the World Bank, which concluded that "doing nothing" to avoid the impacts of climate change would mean a 5–20% reduction in global gross domestic product (GDP) by the middle of the twenty-first century, whereas stabilizing greenhouse-gas concentrations at roughly double preindustrial levels would cost just 1% of global gross domestic product by the middle of this century. A key factor in the calculations is the so-called "discount rate," the rate at which the value of money declines with time. Stated another way, the value one chooses for this economic parameter reflects how much the welfare of future generations is to be compared with future, as yet nonexistent generations.

ORR JC, Fabry VJ, Aumont O, Bopp L, Doney SC, Feely RA, Gnanadesikan A, Gruber N, Ishida A, Joos F, Key RM, Linday K, Maier-Reimer E, Matear R, Monfary P, Mouchet A, Najjar RG, Plattner G-K, Rodgers KB, Sabine CL, Sarmiento JL, Schlitzer R, Slater RD, Totterdell IJ, Weirig M-F, Yamanaka Y, Yool A (2005). Anthropogenic ocean acidification over the twenty-first century and its impact on calcifying organisms. *Nature,* **437**:681–686. According to the authors, today's surface ocean is saturated with respect to calcium carbonate, but increasing atmospheric carbon dioxide concentrations are reducing ocean pH and carbonate ion concentrations, and thus the level of calcium carbonate saturations. They present experimental evidence that key marine organisms—such as corals and some plankton—will have difficulty maintaining their external calcium carbonate skeletons. They used thirteen models of the ocean carbon cycle to assess calcium carbonate saturation patterns under the "business as usual" scenario of the IPCC. In their projections, large portions of the world's oceans will begin to become undersaturated with respect to aragonite, a metastable form of calcium carbonate, by the year 2050. They report that conditions detrimental to oceanic ecosystems could develop within decades, not centuries as suggested previously.

OSTLING RN (2006). Surprise advocate: one of the world's leading biologists calls for 'truce' between science, spirit. *The Spokesman Review,* 29 July 2006, p E3. Ostling interviews Francis S. Collins, past head of the Human Genome Project, about his born-again Christian faith, and the complete compatibility that Collins sees between modern science and a proper reading of the Bible. Collins states that he was raised by nonreligious parents and that he turned into "an obnoxious atheist." But as a medical student he wondered why patients who were suffering and dying retained faith in God. He realized that as a scientist, "you're not supposed to decide something is true until you've looked at the data. And yet I had become an atheist without ever looking at the evidence of whether God exists or not." He began looking and early in the process read the C. S. Lewis classic, "Mere Christianity." Collins has authored a book on the intrinsic compatibility of science and religion, *The Language of God: A Scientist Presents Evidence for Belief.*

PIELKE R, Prins G, Rayner S, Sarewitz D (2007). Lifting the taboo on adaptation. *Nature* **445**:597–598. The authors forcefully argue for immediate international attention to boosting the adaptive capabilities of lesser developed societies already impacted by overpopulation and other factors not directly related to climate change. They give three main arguments for a higher focus on adaptation: (1) the "timescale mismatch," by which they mean that it will take many decades, if ever, for emission-reduction efforts to bring any meaningful climate benefit; (2) the intrinsic vulnerability of human populations to climate-induced events (such as the oft-cited landfall of Katrina in New Orleans) is increasing for reasons that have nothing to do directly with climate change but rather by unsustainable patterns of development; and (3) advocates for the developing nations that have organized on this issue and have put forth the "Delhi Declaration" of 2002, calling for greater attention to adaptation in international climate-change policy negotiations. The authors go on to point out the bizarre distortions in public policy that are being created by the unbalanced focus on climate change, rather than other types of threats to the developing countries. In the Philippines, for instance, sea level rise is occurring at 1–3 mm per year, while excessive ground water extraction is dropping the land surface by over 100 mm per year. Yet the latter issue gets no attention, neither in the media, nor among the thousands of international scientists flying to the latest climate change conference.

RAWLINS W (2007). New map called 'a disservice.' *Raleigh News and Observer,* February 2007, p 14A. It seems surprising that the publication of a plant-hardiness zone map could produce such controversy, but that is exactly what was reported in this newspaper article. The thrust of the article is related to some experts claiming that basing the new map on the most recent fifteen years of data was insufficient, despite the fact that the previous map, published in 1990, was based on only thirteen years of data (from 1974 through 1986). Not surprisingly, the data show a significant northward and inland shift throughout the country for each planting zone. The zones are based on contour map determinations for the average of the fifteen lowest temperatures for each year. The borders for each zone have moved significantly, the precise distance determined in part by local topography, etc. Since this story appeared in a North Carolina paper, that was their focus of attention. Planting zone 7, which corresponds to an average annual low between 0 and 10°F, covers most of the state in the 1990 map but is replaced by planting zone 8 (10–20°F) from the Piedmont region to the south and east in the 2006 version. According to the map, therefore, many plants should now be safe to grow in the Chapel Hill area, such as canna lilies, dahlias, and windmill palms. The "old-timers" don't believe this is true—and it makes for a great news story.

RITTER SK (2007). What can we do with CO_2? *Chemical & Engineering News,* 30 April 2007, pp 11–17. This article highlights several of the innovative ideas that are being tried in order to use carbon dioxide as a direct feedstock for chemical processes. Although the overall concept seems initially compelling, the article concludes on a rather pessimistic note with regard to the real potential for such technologies to have a meaningful impact, "Using CO2 as a raw material probably is never going to reduce atmospheric CO2 levels or lessen the effects of CO2 on climate change – the numbers just don't add up."

RUTTIMANN J (2006). Sick seas. *Nature* **442**:978–80. This article takes a rather alarmist tone in alerting any previously uninformed readers that the increased uptake of carbon dioxide by the oceans has lowered their pH. Because oceans have become more acidic, and will become increasingly so, this will cause the dissolution of calcium carbonate, a key component of the shells of several types of marine organisms. The article summarizes numerous research studies that suggest many forms of marine life are immediately threatened by these changes, putting entire food webs at risk. Consequently, nations that are dependent on obtaining their food from oceans will soon need to develop alternative food sources.

SABINE CL, Feely RA, Gruber N, Key RM, Lee K, Bullister JL, Wanninkhof R, Wong CS, Wallace DWR, Tilbrook B, Millero FJ, Peng T-H,

Kozyr A, Ono T, Rios AF (2004). The oceanic sink for anthropogenic CO_2, *Science,* **305**:367–71. The authors report on an international survey effort during the 1990s that resulted in the first direct measurements of the amount of anthropogenic (man-made) carbon dioxide that has been taken up by the oceans since the extensive burning of fossil fuels began. They conclude that the ocean has taken up approximately 48% of the total emissions, implying that the terrestrial biosphere was a net source during this period. They further conclude that the total carbon dioxide stored in the ocean appears to be about one-third of its long-term potential. They also note that there is already evidence of a positive feedback effect in which the ocean's ability to absorb carbon dioxide is lessened through both chemical and biological effects.

SAMUELSON RJ (2007). Greenhouse simplicities. Newsweek, 20 August 2007, p 47. Samuelson, a regular *Newsweek* contributor, argues against the "Great Conspiracy" theory advocated by the previous week's issue of *Newsweek*. He also takes on the greatest conundrum raised by the global warming issue, which he calls the "great unmentionable"—we lack the technology to reverse current trends and simply don't have a solution for the problem—if it is real. He points out that even with massive regulations, higher energy taxes, huge voluntary conservation measures—as unlikely as he says all of that would be—it would barely cause a blip in the relentless temperature rise. But the final take home message of his piece is to try to correct the flawed premise, as he saw it, of the previous week's issue. Although he acknowledges it is very tempting to treat the debate as a great morality tale, with evil corporations and well-meaning, tree-hugging environmentalists, the issue is vastly more complicated and intractable, and that any dissent about whether the science is truly settled or not should be cheered and encouraged—at least in a free society.

SCHIERMEIER Q (2006). A sea change. *Nature,* **439**:256–60. This is a "news feature" stating that a collapse in ocean currents triggered by global warming could be catastrophic but only now is the Atlantic circulation being monitored. The cause for concern is something called the "global conveyor belt," which transports warm surface water towards the poles and cold deep water back to the tropics. The circulation is driven by differences in temperature and salinity, hence it is also known as the "thermohaline" circulation. This process is particularly important in the North Atlantic, and there is evidence that a "shutdown" of the circulation in this area may have been caused during the sudden collapse of an ice dam in the general area of Hudson Bay several thousand years ago. This would have dumped a sufficiently large amount of fresh water in the ocean to have stopped the circulation for quite some time. It would also have reversed the otherwise general warming trend in northern Europe, which I am sure was not a welcome change for one or two of my Scandinavian ancestors! However, not to worry about this for the next century or so, as one the world's expert on the issue, Wallace Broecker, is quoted here to proclaim, "the notion that a collapse of the thermohaline circulation may trigger a mini ice age is a myth."

SCHIERMEIER Q (2007). What we don't know about climate change. *Nature,* **445**:580–81. In this article, Schiermeier attempts to give a broad and balanced view on the many parts of the underlying science that are not yet settled. The article is placed below an incongruous photograph from Southern California in January 2007 that shows massive icicles clinging to freezing orange trees. At the top of the list of unknowns are the many feedback processes in the models: carbon cycling, ocean currents, decomposing organic matter in the tundra, impacts on oceanic life, etc. The 2007 IPCC report also represented a bit of a retreat when it came to predicting sea level rise, only 9–88 mm by 2100, far less than in the prior reports. However, this is still being "hotly" debated within the climate modeling community, and a new article by Stephan Rahmstorf in *Science,* published online in February 2007 and therefore much too late to influence the mammoth IPCC consensus documents, still predicts a rise of 1.4 m (1,400 mm) by the year 2100. Another burgeoning area of debated science is concerned with climate predictions for particular geographic areas. The 2007 IPCC report was the first to take this on, and many new scientific articles were published on the subject in the year or so prior to the release, and therefore were also excluded from being considered by the United Nations panel of scientists.

SCHROEDER GL (1990) *Genesis and the Big Bang.* New York, NY: Bantam Books, 212 pp. Schroeder is an accomplished theoretical physicist and an Orthodox Jew, who has studied the Hebrew Scriptures and commentaries extensively. In this book and subsequent writings, Schoeder takes the position that any apparent discrepancies between modern science and the Scriptures are merely mirages that vanish upon closer scrutiny. In fact, the book finds amazing degrees of impossibly "coincidental" similarity between the apparent poetry of Genesis 1 and the latest theories of what must have taken place during the first few nanoseconds after the Big Bang.

SEAGER R, Tine M, Held I, Kushnir Y, Lu J, Vecchi G, Huang H-P, Harnik N, Leetma A, Lau N-C, Li C, Velez J, Naik N (2007). Model projections of an imminent transition to a more arid climate in southwestern North America. *Science,* **316**:1181–1184.

SHORT PL (2007). Keeping it clean: the chemical industry is finding ways to secure and expand global water supplies. *Chemical & Engineering News,* 23 April 2007, pp 13–20. The article highlights efforts to meet one of the pressing challenges presented by climate change—a lack of clean drinking water in areas predicted to suffer prolonged droughts. Some of the ideas involve simple technology transfer of proven methods from industrial countries to the developing countries. Other ideas include new separation membrane systems, and some smaller firms have developed self-powered devices using photovoltaic (solar) energy.

SINGER SF (2007). Global warming: man-made or natural? *Imprints,* **36**:1–5. *Imprints* is a publication of Hillsdale College, a relatively small Christian school in Colorado, but with a stated readership over one million. The article is adapted from a lecture delivered on campus by Singer in June 2007.

SMITH TM, Reynolds RW (2004). Improved extended reconstruction of SST (1854–1997). *Journal of Climate,* **17**:2466–2477. The authors are with the US National Climatic Data Center in Asheville, North Carolina, and they present an improved reconstruction of the Sea Surface Temperature (SST) dataset for the past one and a half centuries. Uncertainties in the record are greatest during the nineteenth century (0.4°C) and during the two world wars (0.2°C), but uncertainties are typically on the order of only 0.1°C for the last half of the twentieth century, during which the authors report there has been significant warming, especially since 1970.

SMITH TM, Peterson TC, Lawrimore JH, Reynolds RW (2005). New surface temperature analyses for climate monitoring, *Geophys. Res. Lett.,* **32**,

L14712, doi:10.1029/2005GL023402. In this work, Smith and Reynolds build upon their previous work with the historical sea surface temperature record to add the land surface temperatures and thereby construct global mean temperature records. They provide uncertainty measures that account for both sampling and systematic bias errors and note that the data may now be sampled at finer spatial scales if that is desired. The data extend from 1880 to present, and the authors report that anomalies up until about 1970 all appear to be within the experimental error, but the trend since that time is statistically significant in the direction of global warming.

SOLOMON J (1994). World Views. *Probe Ministries*, http://www.lead-erru.com/orgs/probe/doce/w-views.html. In this piece from 1994, Solomon was one of the first to have attempted a clear definition of a "biblical worldview," although even his spelling was different from that now in common usage. He borrows from another source document (Walsh and Middleton) to come up with the following definition, "A world view provides a model of the world which guides its adherents in the world." He argues that a world view should be rational, supported by evidence, give a comprehensive explanation of reality, and provide a satisfactory basis for living. He goes on to ask six questions of a series of competing worldviews. His questions are: (1) Why is there something rather than nothing? (2) How do you explain human nature? (3) What happens to a person at death? (4) How do you determine what is right and wrong? (5) How do you know that you know? (6) What is the meaning of history? The worldviews discussed by Solomon are the biblical worldview, which he calls "Christian Theism" and six others: Deism, Naturalism, Nihilism, Existentialism, Eastern Pantheism, and New Age or New Consciousness.

SOWELL T (2007). Global warming? or just political hot air? *West Newsmagazine, West Media Inc,* 21 February 2007, pp 44–45. In this article typical of the many responses to news of the IPCC 2007 report, Sowell confronts the theory of man-made global warming as just another of a series of inventions by the political left, both in this country and around the world, who use it as part of their larger agenda to impose big government solutions to all of society's ills. Sowell cites the regular list of skeptics' arguments as to why the theory may not be true and that a little global warming would not be nearly as severe as the activists make it out to be.

STILLMAN D (2007). Zooming in on climate change. *Imaging Notes,* Summer 2007, pp 40–45. Stillman highlights three uses of remote sensing data to monitor climate change impacts: observing ocean plant life and sea surface temperatures, imaging ice, and monitoring carbon sequestration systems.

STEIN LY, Yung YL (2003). Production, isotopic composition, and atmospheric fate of biologically produced nitrous oxide. *Ann. Rev. Earth Planet. Sci.* 31:329–356. This review summarizes the mechanisms that lead to biological production of this important greenhouse gas, and presents the isotopic methods that are being used to develop a global budget of sources and sinks for this material. The article concludes by stating that the biological sources of nitrous oxide far outstrip the industrial sources and that these biological sources are greatly stimulated through the addition of nitrogen fertilizers. The review also states that several uncertainties remain concerning the relative importance of the many biological sources.

THE ROYAL SOCIETY (2005). *Ocean Acidification Due to Increasing Atmospheric Carbon Dioxide,* Policy Document 12/05, June 2005, ISBN 0 85403 617 2, 60 pp.

UCLA AND JPL (2006). UCLA and JPL form partnership to enhance understanding of regional climate change and support future space missions. Press Release, 25 October 2006. This is just one of many examples of novel partnerships that have been announced in recent years as universities and other research institutions vie for the rapidly increasing amount of funding that is being awarded for studies related to climate change.

UNITED STATES CLIMATE CHANGE SCIENCE PROGRAM (2008). *Scientific Assessment of the Effects of Global Change on the United States, A Report of the Committee on Environment and Natural Resources National Science and Technology Council,* Department of Commerce: Washington DC, 271 pp. This comprehensive report provides an extensive review of the current science and the likely effects of additional global warming on the United States. The report draws a similar conclusion to that offered within this book—that current IPCC projections of rate of temperature change are lagging behind the observed rate of accelerating warming.

WADE R (2000). Worldviews, Part 2. *Probe Ministries,* http://www.probe.org. This article is meant both to complement and to provide an update to the piece written by Solomon several years earlier. It defines a new worldview, "postmodernism," that is now running rampant at universities in the United States and elsewhere in the Western World. However, in defining it, Wade makes the observation that it really isn't a worldview, at least not an internally consistent worldview, given that it is based on at least one self-contradictory premise, "absolute truth does not exist," to which we need only ask of the proponents, "are you absolutely sure about that?" Wade goes on to say that postmodernism is really just a "pessimistic mood" about reality, basically "punting" on the whole idea that the universe makes sense, instead asking questions like, "What do I choose to believe is true?"

WEART SR, *The Discovery of Global Warming,* Cambridge, MA: Harvard University Press, 228 pp, 2003. Weart provides readers with a sound summary concerning the history of global warming theory. To a much lesser extent, he analyzes some of its sociopolitical aspects. At the end, he steps out of the scientific domain to strongly advocate for raising gasoline taxes in the United States "by a few dollars" per gallon. He does not attempt to consider the dramatic economic impacts of such a move.

YOO MJ (2007). Planet Action: collaboration to provide climate change tools. *Imaging Notes,* Summer 2007, pp 30–35. The article features an interview with Herve Buchwalter, CEO of the French satellite company, Spot Image. The new collaboration has three aspects: (1) becoming a kind of a "hub" for big science to interact with village projects involving climate change observations; (2) provide a common depository for all of the knowledge created by such projects; and (3) provide materials that raise awareness and support educational programs about climate change.

ZAKARIA F (2007). "Cathedral thinking." *Newsweek,* 20 August 2007, p 48. Zakaria interviews James E. Rogers, chairman and CEO of Duke Energy, about the challenges of improving global energy efficiencies, especially those related to the coal-fired production of electricity. Rogers is not optimistic. One of the telling quotes is the following, "I think we have had chronic underinvestment in energy efficiency. We really need to accelerate that. Mitigation of climate change is not going to happen fast enough. That is the reality." He states that 85% of the incremental carbon dioxide in the next several decades will be coming from developing countries, primarily India and China—countries that were specifically excluded from needing to meet the emission goals of the Kyoto Protocol.

ZAREMBO A, MAUGH TH (2007). Global warming report offers…preview of coming catastrophes: fires, floods and famine, extinctions, droughts and hurricanes, *St. Louis Post Dispatch,* 7 April 2007, p A25. This was the "news" article that accompanied the normal front page graphics that graced many a weekend newspaper of the day with all of the normal scary images: stranded and hapless polar bears looking forlorn on a melting iceberg, world maps with all of the horrendous regional predictions, a third-world woman hoisting a heavy pot of drinking water across an arid and baked lake bed. Here's the lead of the article, "A new global warming report issued Friday by the United Nations paints a near-apocalyptic vision of Earth's future: more than a billion people in need of water, extreme food shortages in Africa, a planetary landscape ravaged by floods and millions of species sentenced to extinction."

Appendix Two

CONTACT INFORMATION FOR THE AUTHOR:

David I Gustafson, Ph.D.
8 The Boulevard Saint Louis, Suite 402
Richmond Heights, MO 63117
phone: 314-727-0795
email: dave@real-whirlwind.org